Hands on Chemistry

Hands on Chemistry

Jeffrey Paradis
Cal State – Sacramento

with
D. Van Dinh
Chad Eller
Holly Garrison
Joel Kelner
Carrie Lopez-Couto
Brian Polk
David Roberts
Kristen Spotz

Boston Burr Ridge, IL Dubuque, IA Madison, WI New York San Francisco St. Louis
Bangkok Bogotá Caracas Kuala Lumpur Lisbon London Madrid Mexico City
Milan Montreal New Delhi Santiago Seoul Singapore Sydney Taipei Toronto

HANDS ON CHEMISTRY

Published by McGraw-Hill, a business unit of The McGraw-Hill Companies, Inc., 1221 Avenue of the Americas, New York, NY 10020.

 This book is printed on recycled, acid-free paper containing 10% postconsumer waste.

1 2 3 4 5 6 7 8 9 0 QPD/QPD 0 9 8 7 6 5

ISBN 0–07–253411–7

Editorial Director: *Kent A. Peterson*
Sponsoring Editor: *Thomas D. Timp*
Developmental Editor: *Joan M. Weber*
Senior Marketing Manager: *Tamara L. Good-Hodge*
Project Manager: *Joyce Watters*
Senior Production Supervisor: *Kara Kudronowicz*
Lead Media Project Manager: *Judi David*
Senior Media Technology Producer: *Jeffry Schmitt*
Senior Designer: *David W. Hash*
Cover Designer: *Emily Feyen*
(USE) Cover Image: ©*Lester Lefkowitz/CORBIS*
Senior Photo Research Coordinator: *John C. Leland*
Supplement Producer: *Brenda A. Ernzen*
Typeface: *10/12 Helvetica*
Printer: *Quebecor World Dubuque, IA*

The credits section for this book begins on page 29-11 and is considered an extension of the copyright page.

www.mhhe.com

Hands On Chemistry
Table of Contents

Hands On Chemistry
Preface: To the Student

Welcome to a new way of doing chemistry laboratory! Our focus in writing this manual has been on developing highly readable experiments that will provide you with a successful learning experience. Our method for developing laboratories begins with identifying concepts that are of particular interest or challenge to students and which we feel would benefit from clarification through laboratory work. From this, objectives are developed which are included in the beginning of each laboratory and which serve as a key focus point for all aspects of the given experiment. The pedagogical approach of the laboratory is then chosen to make the most of the topic we are trying to teach you. For example, some laboratories benefit from a discovery type approach while others are best taught following a more traditional expository approach.

Each experiment contains the sections described below.

- **Title and Author**

The title includes information about the chemistry involved as well as the context of the experiment. We include the author's name on each experiment that they write so that they are given proper credit and so that the students using the laboratory manual can make a connection with the authors. Even though the experiments are edited for overall continuity, each author has a slightly different writing style and we feel this adds to the interest of the manual.

- **Objectives**

The objectives are brief statements outlining the goals for the laboratory. The objectives should answer the questions, "What will the student know after completing the experiment?" and "What will the student be able to do after completing the experiment?" Whenever possible we stress to the students that what they are doing and what they are learning is significant and relates to or enhances material covered in lecture.

- **Introduction**

The introduction consists of a practical or interesting example that serves as a context for the laboratory experiment. This brief attention grabber is intended to make the student want to learn about the material and want to perform the experiment.

- **Background**

The background section is intended to provide the student with the pertinent chemistry required for successful completion of the experiment. The background includes complete descriptions of all relevant chemical equations, explanations of all experimental procedures and information required for analysis of results and data.

- **Overview**

The overview provides the student with the big picture of what they will actually be performing in the experiment and serves to relate the background to the upcoming procedure.

- **Procedure**

The procedure begins with a list of materials (chemicals, glassware and instruments) required for the experiment. The details of the procedure depend, to some degree, on the goals for the experiment. For certain experiments, it is appropriate for the students to develop their own procedure. Some experiments will have provided data tables though usually the students are required to make their own. Some experiments have discovery type components when it helps accomplish the objectives of the experiment. Detailed safety and waste disposal information is also provided to the student for each experiment.

- **Pre-laboratory exercises**

Our goal with the pre-laboratory assignments is to prepare students so that when they enter the laboratory they are fully aware of what they are trying to do and why. After completing the pre-laboratory questions the students should have a firm grasp of the chemistry involved in the upcoming experiment.

- **Post-laboratory work**

The post-laboratory work is our opportunity to make sure that the goals of the experiment have been achieved. Appropriate post-laboratory work involves writing up the laboratory experiment, performing calculations using data from the experiment, answering open-ended writing assignments and performing analysis of data and errors.

What's the Matter? The Nature of the World Around Us

Jeffrey Paradis

OBJECTIVES

- Explain various daily observations in terms of the particulate nature of matter.

INTRODUCTION

The theory that matter is made up of atoms is fundamental to our current understanding of chemistry (as well as aspects of biology, physics, astronomy, engineering and geology). Until recently however, scientists had only indirect evidence for the existence of atoms. Being able to see atoms would allow scientists to, among other things, design better computer chips and drugs. You might wonder why we can't just make a microscope powerful enough to look at atoms. It turns out that the wavelength of visible light is too large to resolve the distances between individual atoms. This is analogous to trying to read Braille while wearing a baseball glove.

In the early 1980's scientists at IBM developed a method for making images of the surface atoms of certain materials. They named this method Scanning Tunneling Microscopy (STM) and were awarded the 1986 Nobel Prize for their ground-breaking work. Never before had scientists been able to enter the atomic world with such ease and clarity (Figure 1).

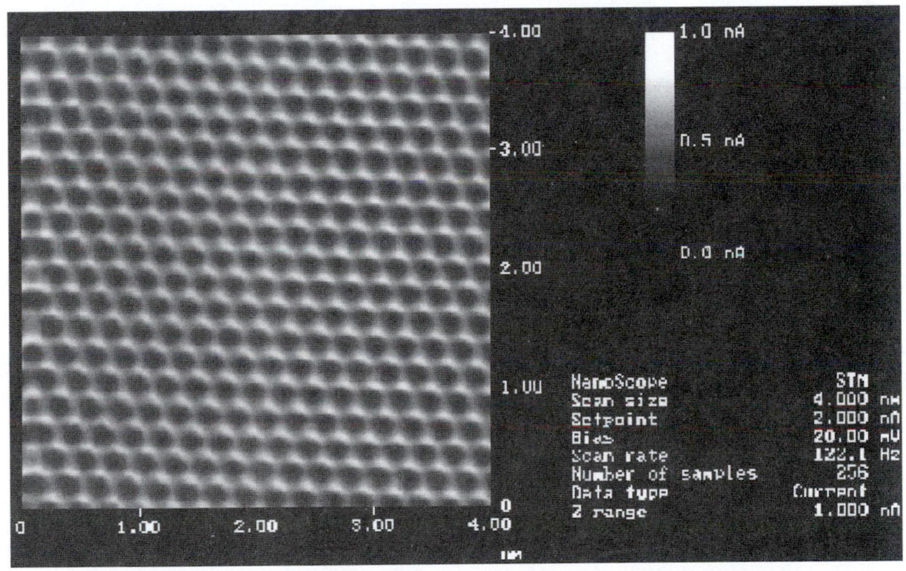

Figure 1. An STM image of the surface of graphite (carbon).

BACKGROUND

Matter is Particulate

Chemistry is often termed "the central science" because it involves the study of matter. Our study of chemistry will therefore benefit if we have a firm conceptual understanding of the nature of matter. Dalton's first postulate provides us with some insight. The postulate states that "An element is composed of extremely small indivisible particles called atoms". For our purposes we can expand on Dalton's statement and think of all matter as being made up of incredibly small particles. This postulate is in conflict with the common misconception people have that matter is continuous. In studies of elementary to university students, over half of the students held concepts suggesting a perception of matter as a continuous medium, rather than as an aggregation of particles. This misconception is so common because our day-to-day experience with matter seems to suggest that it is continuous (for example, we can't see the individual particles that make up our desk). It is the goal of this experiment that you will gain an appreciation of matter as being made up of infinitely small particles. This knowledge will serve as a foundation for all later learning in chemistry.

OVERVIEW

In this experiment, you will examine four different everyday phenomena from the point of view of matter being composed of extremely small particles (atoms, ions and molecules) that are too small to see.

PROCEDURE

Part A: Evaporation of Alcohol

Chemicals Used	Materials Used
Rubbing alcohol	Chalk board Cotton ball

Take one cotton ball and soak it with alcohol so that it is thoroughly wet but not dripping. Drag the cotton ball across the chalkboard to make a wet streak about one foot long. Describe what happens to the streak. Does it happen all at once? Approximately how long does it take to happen?

Part B. Dissolving Crystals

Chemicals Used	Materials Used
Container of Kool-Aid crystals (dark color)	100 mL graduated cylinder Spatula

Fill the graduated cylinder with fresh water. Let it sit for a minute so any air bubbles may rise and the solution is clear. Prop a piece of white paper up behind the graduated cylinder so it will be easier to make your observations. Get some Kool-Aid on the end of the spatula – no more than the size of a pea. Drop the crystals into the center of the top of the water in the graduated cylinder. Do NOT stir. In addition to a written description of what you see happening, include a sketch of what you observe a few seconds after the colored crystals are added to the water. After you have written your observations empty the graduated cylinder into the sink. Do not drink the Kool-Aid.

Part C. Mystery balloon

Materials Used
A blown up balloon with a few drops of scented oil in it

Carefully pick up the balloon and smell it. Other than the smell of the latex balloon itself, record what else you can smell.

Part D. Mixing liquids

Chemicals Used	Materials Used
Bottle of alcohol (colored yellow) Bottle of water (colored blue)	Glass stirring rod Graduated cylinders (2)

Pour approximately 20.0 mL of water into one graduated cylinder and approximately 20.0 mL of alcohol into the other graduated cylinder. Record the volumes of each liquid. In this experiment you will be watching for slight changes in volume so make your readings of volume as carefully as possible. Carefully pour the water into the alcohol so none is lost and stir carefully to mix the two liquids. Record the volume and color of the mixture. When you are done, pour the "water + alcohol" down the sink.

What's the Matter? The Nature of the World Around Us

Name:	Lab Instructor:
Date:	Lab Section:

PRE-LABORATORY EXERCISES

1. Define the underlined terms in the BACKGROUND section.

2. Give three examples of things that are samples of matter. Give three examples of things that are not samples of matter.

3. The scale on the image of the carbon atoms (INTRODUCTION, Figure 1) indicates 4.00 nm. At this point you are not expected to know what that means, but it translates to 0.000000004 meters. Knowing this, roughly what is the radius, in meters, of a single carbon atom according to the STM image? Show your work.

4. The head of a pin is as big as about 600,000,000,000,000,000,000 atoms of aluminum! If every person on the Earth counted one aluminum atom every second, how many years would it take us all to count all of the aluminum atoms on the head of the pin? Comment on what the magnitude of this number tells us about the size of an atom.

What's the Matter? The Nature of the World Around Us

Name:	**Lab Instructor:**
Date:	**Lab Section:**

RESULTS and POST-LABORATORY QUESTIONS

Part A: Evaporation of Alcohol
Describe your observations:

How does your observation support the idea that matter is made up of extremely small particles that are too small to see? How might the evaporation process look different if matter was continuous (not made up of tiny particles)?

Part B. Dissolving Crystals
Describe and draw a picture of what you observed:

How do your observations support the idea that matter is made up of extremely small particles that are too small to see? How might the process of dissolving look different if matter was continuous (not made up of tiny particles)?

As the crystals fell through the water what caused the water to change color?

OVER →

RESULTS and POST-LABORATORY QUESTIONS continued...

Part C. Mystery balloon

What other smell was present?

Propose an explanation for the fact that a chemical inside of the balloon can be detected outside of the balloon.

Part D. Mixing liquids

Volume of water (blue):

Volume of alcohol (yellow):

Actual volume of "water + alcohol":

Color of the "water + alcohol":

What was the expected volume of the "water + alcohol"?

Try to come up with at least 2 possible explanations for your observation. How could you go about testing each possible hypothesis? Try to explain why the final volume and the final color was different than the sum of the two liquids that made it up.

As a model for helping us understand what occurred when mixing the water and alcohol, imagine two graduated cylinders one with approximately 20 mL of large ball-bearings and the other with approximately 20 mL of small ball-bearings. Next imagine you pour the graduated cylinder of small ball-bearings into the cylinder of large bearings. Predict what you would see happen. How does this model help you understand/revise what was happening earlier when you mixed the two liquids. Explain.

RESULTS and POST-LABORATORY QUESTIONS continued...

1. What is between the atoms in a sample of matter? Which part of this laboratory helps support your answer?

2. Using your knowledge about the particulate nature of matter, propose a description for what is happening when you use a tea bag to make a cup of tea.

3. Do you think that foods would have a smell if matter wasn't particulate? Explain.

The Properties of Some Common Metals: Why NASA Needs Chemistry

Holly Garrison

OBJECTIVES

- Determine chemical and physical properties of various substances.
- Reach a conclusion based on evaluating the merits of several variables.
- Design a laboratory experiment to determine the density of a metal.

INTRODUCTION

On January 28, 1986, the Space Shuttle Challenger was launched for the last time (Figure 1). A burst of white following the ignition of hydrogen and oxygen gases could be seen. This tragic explosion took the lives of seven astronauts. Subsequent investigations revealed that the cause of the accident was the failure of a pressure seal in one of the shuttle's solid rocket boosters. This was due to the vaporization of an o-ring seal (a synthetic rubber band). It was known that the o-rings were less effective in forming seals at temperatures below 53°F. The shuttle was launched, regardless of this fact, at a temperature just above freezing.

Figure 1. The explosion of the Space Shuttle Challenger

This great loss vividly illustrates the crucial relationship between an understanding of basic science and the world of high technology. Researchers and engineers must be knowledgeable about the properties of the substances they are working with and must be able to communicate effectively as teams of professionals. For this lab, you will be conducting experiments to determine the properties of several metals in order to select the one best suited for a theoretical mission to the planet Venus.

BACKGROUND

Physical and Chemical Properties

All elements and compounds can be classified according to their physical and chemical properties. Scientists can use properties to determine the identity of unknown materials. If local fish and wildlife are getting sick, scientists might run a series of test on the local water to determine something about the properties of the materials dissolved in the water. One property alone might not be enough, but a complete list of properties is just like a fingerprint for a chemical. Imagine that someone you have never met before is going to meet you at the airport. Knowing that the person is a woman wouldn't be that helpful. But, if you know she is blond, short, wears glasses and has a green shirt on you will probably find her. Likewise, simply knowing that the boiling point of a certain liquid is 127°C probably won't allow you to identify the liquid since many liquids have the same boiling point. But, if you also know the density and freezing point of the liquid, you are well on your way to identifying the unknown.

Some examples of physical properties include color, boiling point, melting point, density, thermal conductivity, and specific heat. The last three are of importance in our investigation. Let's look at them briefly. Density is a temperature dependent intensive property. The greater an element's mass per unit volume, the higher its density. For example, a piece of lead is denser than a piece of tin. Thermal conductivity deals with how quickly heat is carried through a material. For a practical example of how thermal conductivity works, go outside and sit on some metal bleachers and then on a plastic or cloth seat. Both materials are at the same temperature but the metal feels hotter than the cloth because the metal is a better thermal conductor. Water is said to have a high specific heat capacity, because it takes an amazingly large amount of heat to bring about changes in its temperature due to the presence of hydrogen bonds. You will learn much more about this later in your chemistry course.

Chemical properties include flammability, corrosiveness, and reactivity with acids. For example, some metals are not oxidized by acids and will remain unchanged upon exposure. Other metals slowly bubble as gases are given off (Figure 2). Still other metals react explosively when placed in contact with the acid. In your study, you will be observing how different metals react with hydrochloric acid.

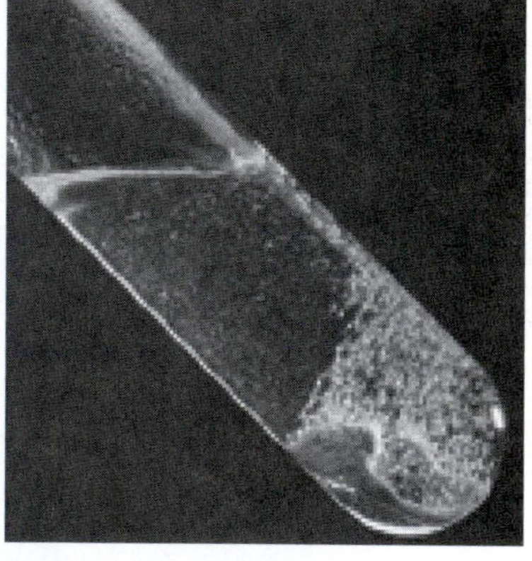

Figure 2. Reaction of nickel with HCl

The Scientific Method

The scientific method is an approach to problem solving. Scientists have used the scientific method to conduct experiments for ages. In addition, YOU use the scientific method informally everyday when you try to figure out why, for example, your car won't start. From the statement of the question to be investigated ("Why won't my car start?") comes a possible explanation or hypothesis ("The battery is dead"), followed by steps to see if that explanation is a plausible one ("No, the car lights come on, so the battery is OK. Maybe the car is out of gas…"). Procedures and results are documented so that other scientists can recreate the same experiments. You will be expected to carry out your investigations based on this model, and to write up your experimental procedure so that others can reproduce your experiments and results.

Mission to Venus

For this lab you are in charge of developing a landing probe that will visit the surface of the planet Venus (Figure 3). Before designing the probe, you'll need to know what you are up against. The conditions on the surface of Venus need to be taken into account when constructing this spacecraft. Venus is a relatively small planet, closer to the sun than the Earth is, with a fairly consistent temperature of 475°C. The atmosphere of Venus is thick and is composed of 96.5% carbon dioxide. This large amount of a "greenhouse gas" maintains the high temperature on Venus, even at night. The pressure on the surface is similar to that placed on a submarine 3000 feet deep in the Atlantic Ocean. There are constant storms on Venus, with winds blowing faster than in any hurricane on Earth. Lightning fills the sky of this dry planet. The clouds surrounding Venus are made of droplets of acid that are strong enough to dissolve most metals. Many probes have been destroyed on missions to this forbidding planet. These conditions limit which materials can be used in the probe. Knowing about the properties of different metals will enable you to make an informed decision.

Figure 3. Venus

OVERVIEW

In this experiment, you will use information about the surface of Venus and the results of experiments that you perform to choose a metal appropriate for constructing a space probe. You will investigate the density, relative thermal conductivity, reactivity with acid and the specific heat capacity of various metals and alloys.

PROCEDURE: You will be provided with samples of various metals and alloys. Choose 3 metal or alloys to analyze in Parts A-D.

Part A. Reaction with Hydrochloric Acid

Chemicals Used	Materials Used
Various metals (1-cm long wire) Hydrochloric acid (2M)	Plastic pipet Well plate with 6 or 12 wells Steel wool

CAUTION: 2 M hydrochloric acid is caustic. If any is spilled on your skin, immediately rinse with running water and inform your laboratory instructor. As always, you should wear your goggles at all times when working in the laboratory

Clean a 1-cm long piece of metal wire with steel wool. Place the metal wire in an empty well of your well plate. Using a plastic, disposable pipet, half-fill the well containing the metal pieces with hydrochloric acid. Record your observations. Repeat with the other two metals of your choice. The acid and metal should be discarded according to instructor directions.

Part B. Specific Heat Capacity

Chemicals Used	Materials Used
De-ionized water Various metals (cubes or 30 cm long wires)	Balance 250-mL beaker Graduated cylinders Hot plate Styrofoam cups nested in beakers (3) Thermometer Tongs

CAUTION: Be careful, hot plates can stay warm for a long time after they are turned off.

Obtain a cube or wire of each of the three metals you chose for analysis in Part A. The three pieces should be of approximately equal mass. Place the metal cubes or rolled up wires into a 250-mL beaker with 100 mL of water. Heat the water and metals to 100°C. Using tongs, remove the pieces of metal and place them in separate Styrofoam cups containing about 20 mL of room temperature water (or enough water to just cover the metal). Record the maximum water temperature. Dry off the cubes or wires and return them to be reused.

Part C. Thermal Conductivity

Chemicals Used	Materials Used
De-ionized water Various metal wire (about 15 cm long) Butter, kept on ice (about 1 cm^3) Macaroni (3 pieces)	Various beakers 50-mL graduated cylinder Hot plate Thermometer Tongs or Hot mitts Spatula

CAUTION: Be careful, hot plates can stay warm for a long time after they are turned off.

For each metal (again choose the same three metals as in Parts A and B) take a length of wire and use butter to attach pieces of macaroni to the ends. Bring 50 mL of water to a boil in a beaker appropriate for the length of your wire. Take the boiling water off the heat source and place it on a paper towel to catch the melting butter. Place the wires into the beaker of hot water (Figure 4). Using a stopwatch, time how long it takes the butter to melt and the macaroni to fall from the end of the wire. Repeat the procedure twice.

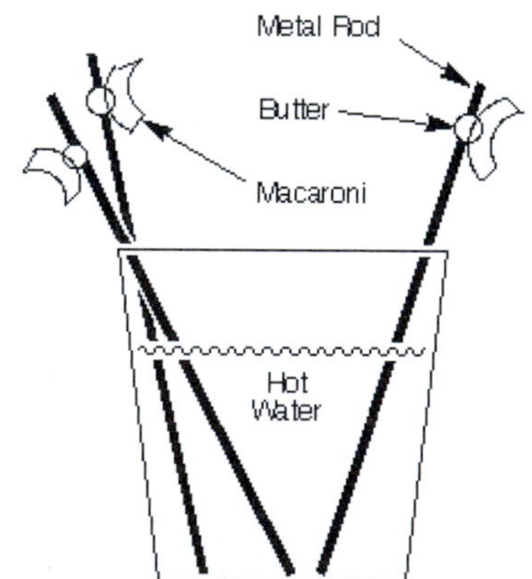

Figure 4. Set-up for measurement of thermal conductivity

Part D. Density

Chemicals Used	Suggested Materials
De-ionized water Various metals (cubes or wires)	Balance Various beakers Various graduated cylinders Rulers

Follow the protocol you wrote in your pre-lab work to determine the density of each metal.

An Investigation of the Properties of Some Common Metals: Why NASA Needs Chemistry

Name:	Lab Instructor:
Date:	Lab Section:

PRE-LABORATORY EXERCISES

1. Define the underlined words in the BACKGROUND section.

2. A 1.50 cm^3 sample of "Metal X" weighs 2.65 grams. A 1.00 cm^3 sample of "Metal Y" weighs 1.78 grams. Which metal is denser? Explain.

3. List three properties of the material used to make a space probe to Venus. Discuss why each of these properties is important and indicate whether each is a chemical or physical property and an intensive or extensive property.

OVER →

PRE-LABORATORY EXERCISES continued...

4. Write a procedure to determine the density of a piece of metal wire. Follow the model used in this experiment. Include a list of chemicals and materials. Be sure to copy the procedure into your laboratory notebook so that even after you turn in your PRE-LAB EXERCISES, you will have the procedure available while performing the experiment.

An Investigation of the Properties of Some Common Metals: Why NASA Needs Chemistry

Name:	Lab Instructor:
Date:	Lab Section:

RESULTS and POST-LABORATORY QUESTIONS

Part A. Reaction with Hydrochloric Acid
Make a table summarizing your observations for each metal.

Part B. Specific Heat Capacity

Metal	Mass of Metal	Change in water temperature (final temp. – initial temp.)

Why was it important that the pieces of metal have similar masses?

Which of the three metals had the highest specific heat capacity? Explain.

Part C. Thermal Conductivity

	Time Elapsed			
Metal	Trial 1	Trial 2	Trial 3	average

Describe a different method that could be used to improve the accuracy and precision of the data collected in the thermal conductivity experiment.

OVER →

Part D. Density

Based on the experiment you developed for density, design an appropriate table for recording your data. Show your calculations, taking into account significant figures. Report your densities in both g/mL and lb/in^3.

1. Use the Internet or a text-book to find the melting points of the three metals you used in this experiment. As always, your answers must include relevant units. A good web site found by searching for "periodic table" will typically have this type of information listed for each element. Record the URL of the web site where you found the information along with the date when you accessed the site. For bonus work, your instructor may also have you find the price of each of the three metals you used.

2. On a separate piece of paper, write an essay answering the following question: Using your experimental data (Parts A – D) and the information you found on melting point (and price?) from question #1 above, present your choice of metal for the space probe. Which properties are most pertinent to your decision? You may have found in your investigation that there was no clear "winner." One of the metals may have appeared to be the best choice based on a certain characteristic, but was not nearly as good in regards to another. You should make your selection depending upon what you feel is the best combination of properties. Use your specific results to defend your choice.

Alchemy and the Origins of Modern Chemistry: All that Glitters Isn't Gold

Brian Polk

OBJECTIVES

- Develop observational skills while performing a series of chemical reactions.
- Practice using various laboratory equipment.
- Gain exposure to basic laboratory safety issues.
- Change copper into gold and find the answer to the meaning of life…

INTRODUCTION

Alchemy has been described as a mixture of science, medicine, magic and religion. Early alchemists believed that by changing less valuable metals into gold, they could make a major step towards everlasting life. Though much of the tradition of alchemy has been discredited today, alchemists played a major role in the advancement of science. Alchemists discovered many substances used in everyday life like alcohol, hydrogen, phosphorus and gunpowder. They also developed the scientific application of processes like distillation, evaporation, extraction and filtration. Figure 1 shows a portion of a painting of an alchemist at work.

Figure 1. A portion of a painting by Joseph Wright entitled, "The Alchemist in search of the Philosopher's Stone, Discovers Phosphorus, and Prays for the successful Conclusion of his Operation as was the custom of the Ancient Cymical Astrologers".

BACKGROUND

A Parchment in the Attic

Imagine you are rummaging through your family's attic and you find an old piece of parchment sandwiched between several books. The parchment is covered with strange symbols like those shown in Equations 1a and 1b.

♈ + ♊ → ♑ Equation 1a

♑ + Δ → ♓ Equation 1b

The equations strike you as resembling chemical reactions. You had heard that one of your ancestors was an alchemist, but never gave it much thought, until now. Further exploration uncovers the key to the mystery (Table 1). Your hunch was correct, the parchment does contain chemical equations and if your translation is correct, it seems your ancestor was pursuing a method of changing copper into gold!

Table 1. Key used to decipher chemical processes described in Equations 1a and 1b.

Symbol	Element
Silver	♑
Gold	♓
Copper	♈
Zinc	♊
Heat	Δ

Though you assume it probably won't work (or else your family would have been a lot richer by now) you decide it is worth a try and you bring the parchment to school to show your chemistry professor. Your professor suggests it would be a great exercise for you to replicate the work and analyze the results (and that's just what you'll do in this experiment).

The More Things Change...

Before you embark on the road to riches, however, you should know the difference between physical and chemical changes. Physical changes occur with no change in chemical composition. For example, when an ice cube melts, the order and arrangement of the water molecules changes, but the same chemical is still present before and after the change. By contrast, chemical changes (also called chemical reactions) accompany a process in which one or more substances are converted into other chemical substances. When burning gasoline, the octane in the gasoline undergoes a chemical change resulting in the formation of chemicals that were not originally present (carbon dioxide and water). It is important to note that although chemical reactions result in new substances, the same constituent elements will still be present throughout the process.

OVERVIEW

In this experiment you will take a penny and transform it according to the series of steps from the alchemist's parchment into what appears to be silver and then gold. As scientists, however, we must validate this finding. We can do this by finding the density of our "gold" and comparing that to the known density of gold.

PROCEDURE: Though you will be performing the reactions on your own penny, pair up with another student to share the beakers of chemicals and the space in the fume hood.

Chemicals	Materials
Zinc filings	125-mL Beaker
NaOH (3 M, aq)	Tongs
HNO_3 acid bath (3 M, aq) in hood	Bunsen burner
Clear nail polish	Hot plate
Pennies (pre-1982) **	Ruler
	Analytical balance

Caution: The entire procedure should be completed **in a fume hood**. Both 3M HNO_3 and NaOH are caustic and should be handled with care. If any is spilled on your skin, immediately rinse the exposed area under running water and inform your laboratory instructor. As always, wear goggles at all times when working in the laboratory.

1. *Clean your penny*: Working in the fume hook, use tongs to dip a pre-1982 penny into a beaker containing the HNO_3 acid bath for 2-3 seconds. Rinse the penny with de-ionized water and dry it. Record the diameter, mass and appearance of the "clean" penny.

2. *Changing copper to silver*: Place about 1.5 grams of zinc into a clean beaker and add enough 3 M NaOH to completely cover the zinc. Heat the beaker on a hot plate in the hood. When steam begins to form, drop the clean penny into the NaOH solution. When bubbles of gas can be seen escaping, carefully remove the coin from the solution with tongs. Turn off the hot plate. Rinse the penny with de-ionized water and dry it. Record the diameter, mass and appearance of the "silver" penny.

3. *Changing silver to gold*: Take the coin again with your tongs and heat it gently in the outer core of a Bunsen burner flame. A change should occur quite suddenly. Stop heating a few seconds after this change occurs. Cool the hot penny by holding it under running water. Again, record the diameter, mass and appearance of the "gold" penny.

4. Pool your pennies with several other groups and measure the thickness of a column of 10 pennies.

5. Preserve your coin by painting it with finger nail polish. If you wish, repeat steps 1 and 2 and preserve the penny with nail polish after the first change.

6. Dispose of all solutions according to your instructor's directions.

** Before 1982, pennies were made primarily of copper (95%). After 1982 however, the composition of "copper" pennies became over 97% zinc with a copper coating.

Alchemy and the Origins of Modern Chemistry: All that Glitters Isn't Gold

Name:	Lab Instructor:
Date:	Lab Section:

PRE-LABORATORY EXERCISES

1. Make a table summarizing the symbol, atomic mass, atomic number and density for copper, silver and gold.

2. Rewrite Equation 1a and 1b from the BACKGROUND section using current chemical symbols.

3. Are the following examples chemical or physical changes? Explain.
   a. Baking a cake

   b. Shredding a piece of paper

OVER →

PRE-LABORATORY EXERCISES continued...

4. In this experiment, you will need to find the volume of a penny. As practice, consider the following question. A small, short cylinder has a diameter of 9.8 cm and a height (thickness) of 3.5 cm. Draw a picture of the cylinder and label the diameter, the radius and the height. Calculate the density of the cylinder if the cylinder has a mass of 853 g. The volume of a cylinder can be found from the following equation: $V=\pi r^2 h$ where r = radius and h = height.

5. Summarize the safety precautions that must be taken to avoid injury during each step of the procedure.

Alchemy and the Origins of Modern Chemistry: All that Glitters Isn't Gold

Name:	**Lab Instructor:**
Date:	**Lab Section:**

RESULTS and POST-LABORATORY QUESTIONS

1. Make a table summarizing the diameter, the mass and the appearance for each of the three pennies ("cleaned copper", "silver" and "gold").

2. Calculate the volume of each of each of the three pennies, using the average thickness that you found from the thickness of the 10 pennies and the equation used in the pre-laboratory question 4. Calculate the experimental density of each of the three coins. Show all your work. Make a table summarizing the volume and experimental density of each of the three pennies.

OVER →

RESULTS and POST-LABORATORY QUESTIONS continued...

3. Do you think the final coin is really made of gold? Justify your answer using your experimental results and your basic knowledge of chemistry. Your explanation should include a comparison with the known densities of copper, silver and gold (pre-laboratory question 1) and a statement about whether it is possible for elements to change during a chemical reaction.

Classification of Matter: Basic Separation Techniques

Kristen Spotz

OBJECTIVES

- Differentiate between elements, compounds, solutions (homogeneous mixtures) and mixtures (heterogeneous mixtures).
- Practice various physical separation techniques.
- Develop an experimental protocol to separate the components of a mixture.

INTRODUCTION

Life as we know it would not be possible without the petroleum industry. At oil refineries, chemists and engineers take crude oil and separate it into useful components such as gasoline, heating oil, kerosene, starting materials for plastics and pharmaceuticals, diesel fuel, lubricants, waxes and asphalt. The physical separation of the components of crude oil is achieved by a method know as fractional distillation. In fractional distillation, the material being separated (in this case, crude oil) is heated and the components that are the most volatile are the first to vaporize. The molecules that escape from the liquid move up the tall distillation tower (Figure 1). The components that have the lowest boiling point reach the top of the tower where they are condensed and collected in much purer form. The components with higher boiling points are collected at the higher temperatures found at the lower part of the tower. Next time you see the tall towers of an oil refinery, think about the physical separations that are occurring inside and try to imagine what your life would be like without fractional distillation.

Figure 1. The distillation tower of an oil refinery

BACKGROUND

Take a look around you. All matter exists as an <u>element</u>, a <u>compound</u> or a <u>mixture</u>. Mixtures are by far the most common form of matter on earth. Crude oil, drinking water, soil, your body and the air you breathe are all examples of mixtures.

Mixtures are further classified into two broad types: homogeneous and heterogeneous. Heterogeneous mixtures are sometimes simply called "mixtures" and have a variable composition with observable boundaries between the components. For example, cereal mixed with milk is a heterogeneous mixture because the composition varies and there are distinct observable boundaries between the components. Homogeneous mixtures are sometimes called "solutions". Solutions have a uniform composition and no visible boundaries between the components. For example, air is a homogeneous mixture because the various gaseous molecules (CO_2, N_2, O_2, Ar...) are uniformly intermingled on the molecular level.

The flowchart Figure 2 provides a useful method for classifying matter. For example, our bowl of cereal and milk is not uniform throughout and is therefore a heterogeneous mixture.

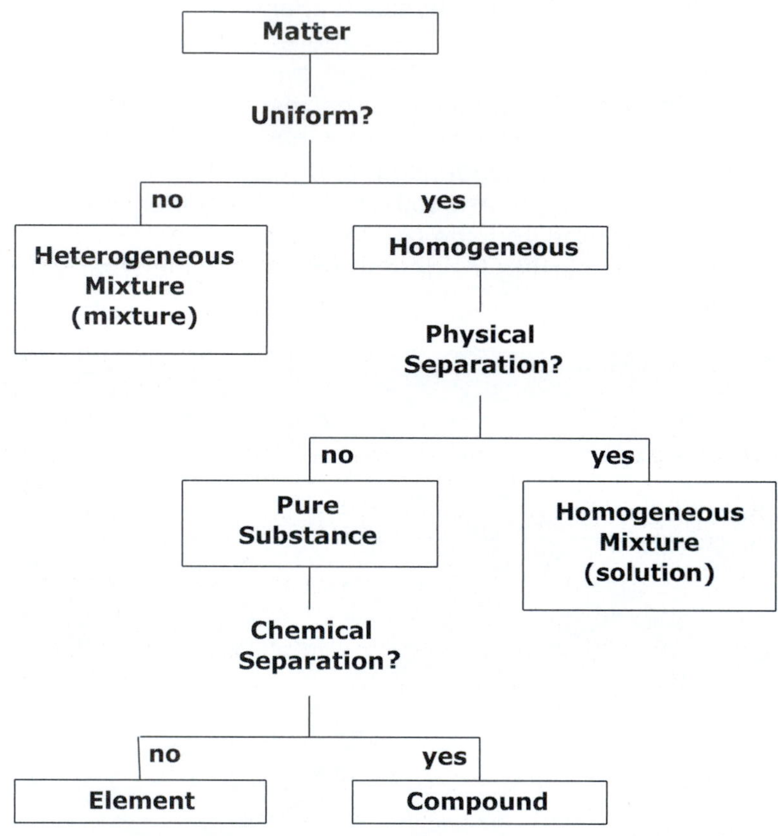

Figure 2. Flow chart for determining if a sample of matter is a mixture, solution, element or compound.

Separating the Components of a Mixture

To fully use the flow chart in Figure 2, it is necessary to discuss what is meant by physical and chemical separations. Physical separations depend on the physical property of the substances in the mixture and occur without any chemical changes. The fractional distillation described in the INTRODUCTION is an example of a physical separation because the chemical components of the crude oil have not been changed; they have only been isolated from each other. The separated components could, in fact, be mixed together to give back the crude oil. Examples of physical separation methods include chromatography and filtration. By contrast, chemical separations depend on chemical properties and result in new chemical compounds and elements that were not originally present. An example of a chemical method of separation is electrolysis, which involves the use of electricity to separate the individual elements of a compound.

The components of both heterogeneous and homogeneous mixtures can be separated by physical methods. If the matter under study cannot be separated by physical methods then the matter exists as a pure substance. The nature of the pure substance (element or compound) can be determined based on whether a chemical separation can then be used to further separate the substance. Elements cannot be further broken down by chemical methods, while compounds can be further separated into their constituent elements.

As an example, oxygen gas can be analyzed using the flow chart in Figure 2. Oxygen gas has a uniform composition with no distinguishable parts and is therefore homogeneous. Because oxygen can not be separated by physical means, oxygen gas must be a pure substance. Oxygen is already composed of only one element and so it can not be further separated by chemical methods. According to our flow chart, oxygen gas is defined as an element.

Some Methods of Physical Separation

We will now look in more detail at some of the methods of physical separation that will be used in this laboratory experiment.

Chromatography: Chromatography is the general term applied to a family of separation techniques that involves two phases, the stationary phase (the phase that stays still) and the mobile phase (the phase that moves during the experiment). The stationary phase is typically a solid and the mobile phase can either be a liquid or a gas. The basis of chromatography lies in the components of the mixture having different affinities (attraction) for the two phases in the chromatography system. For example, the component in the mixture that has a high affinity for the mobile phase moves relatively quickly through the system as it is pulled along with the mobile phase. By contrast, a component of a mixture that has a high affinity for the stationary phase moves relatively slowly. In this experiment, we will be exploring a simple type of chromatography known as paper chromatography.

Filtration: The basis of filtration lies in the differences in size of the particles being separated. Filtration is commonly used to separate a liquid (individual molecules) from a solid (larger particles). Typically, smaller particles pass through the funnel under the force of gravity or a vacuum while the larger particles remain captured in the filter.

Sublimation: Because certain substances have the ability to easily sublime, whereas others do not, <u>sublimation</u> can be used as a separation technique. For example, suppose you were asked to separate a mixture of sugar and iodine. Knowing that iodine has the ability to sublime, you can heat the mixture in a closed system causing the iodine to vaporize. The gaseous iodine can then be condensed using ice, while the sugar is left behind.

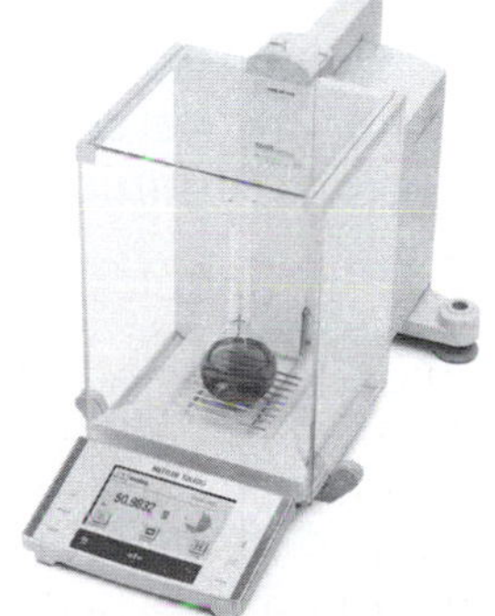

Figure 3. Sublimation of iodine.
When the I_2 comes in contact with the cold watch-glass, sold crystals of I_2 are deposited.

OVERVIEW

You will begin by classifying a variety of samples as elements, compounds, mixtures or solutions. From these samples, you will then run tests to determine whether your classification of ink and water was correct. Finally, you will create your own protocol for separating the four components of a mixture. The entire experiment will be carried out in pairs unless you are instructed to do otherwise.

PROCEDURE

Part A. Classification of Matter

Chemicals Used	
De-ionized water	Copper wire
Hydrochloric acid (aq)	Aluminum can
Sodium chloride (s)	Rock
Sodium chloride (aq)	Salad dressing
Mercury (II) oxide (s)	Graphite (in a pencil)
Nickel (II) nitrate (aq)	Ink (colored markers)
Carbonated soft drink	

Your instructor has set up various labeled stations around the laboratory. For each sample determine whether the matter is an example of an element, compound, mixture or solution. Be able to justify your answer based on your observations.

Part B. Separation Methods: Verifying the Classification of Water and Ink

Chemicals Used	Materials Used
De-ionized water	Filter paper
Ink (colored markers)	100-mL Beaker
Hydrochloric (2 M, aq)	9-Volt Battery
	Tongs

1. Electrolysis: Place the battery up-right in the bottom of the beaker. Fill the beaker with de-ionized water. If nothing happens, add 3-4 drops of 2 M HCl (aq). Record your observations. Using the tongs, carefully remove the battery from the water. Rinse the battery with de-ionized water and make sure that it is thoroughly dry. Clean and dry the beaker. Alternatively, this part may be done as a demonstration by your instructor.

2. Paper chromatography: Cut a piece of filter paper into a 0.5-inch wide strip. Using a colored magic marker put a dot about 2 cm from the bottom of the strip of filter paper. Place the strip vertically into a beaker with a thin layer of water in the bottom. Make sure the water does not cover the dot. After 15 minutes, record your observations and save your filter paper to tape to your lab report. Repeat with a different color ink.

Part C. Methods of Physical Separation: Devising your own Protocol

Chemicals Used	Materials Used
Iodine, I_2 (s) "Salt", Sodium chloride, $NaCl$ (s) "Sand", Silicon dioxide, SiO_2 (s) Iron, Fe (s)	100-mL Beaker Hot Plate or bunsen burner Analytical balance Metal spatula Watch glass Vacuum flask and tubing Filter paper Buchner funnel and plastic ring Fume hood Magnet De-ionized water Ice Heat protecting gloves

CAUTION: The sublimation of iodine must be carried out in the fume hood. Iodine stains hands and clothing. Be careful, hot plates can stay warm for a long time after being turned off. Use the heat protecting gloves when handling hot glassware. Your instructor will show you how to use the vacuum flask and tubing.

1. Create a protocol using the "**Materials Used**" to separate a mixture containing the four "**Chemicals Used**". Show your instructor your procedure before beginning. If you are using the vacuum filtration flask, ask your instructor for a demonstration.

2. To a tarred weighing paper, add pea-sized scoops of silicon dioxide, iron, iodine and sodium chloride. Record the mass after each addition. Gently stir the components.

3. Use the protocol you have written to separate the four components of your mixture. Attempt to get the cleanest separation possible. As you separate out each component, record its mass.

Classification of Matter: Basic Separation Techniques

Name:	Lab Instructor:
Date:	Lab Section:

PRE-LABORATORY EXERCISES

1. Define the underlined words in the **BACKGROUND** section.

2. What is the key difference between an element and a compound?

3. Can the relative amounts of the components of a mixture vary? Can the relative amounts of the components of a compound vary? Explain.

4. The tap water in your home leaves white deposits after evaporating. Does the evaporation involve a physical or chemical separation? Is your tap water a mixture, a solution, a compound or an element? Explain.

OVER →

PRE-LABORATORY EXERCISES continued...

5. Describe an example of a physical method that could be used to separate a pile of sugar and sand.

Classification of Matter: Basic Separation Techniques

Name:	Lab Instructor:
Date:	Lab Section:

RESULTS and POST-LABORATORY QUESTIONS

1. On a separate piece of paper, construct a table summarizing whether each sample of matter in **Part A** is an example of an element, compound, mixture or solution. Explain your answers.

2. Did the tests you performed in **Part B** confirm or disprove your initial ideas about the nature of water and ink? Explain.

3. Tape one of your filter papers with the separated components of ink in the space below. What color ink did you originally use? Label the individual colors that have been separated. Could these components be recombined to give back the original color of ink? Explain.

4. For each separation step you performed in **Part C**, name the property being used as the basis for each separation, state whether each separation is physical or chemical and list the components of the mixture still present.

OVER →

RESULTS and POST-LABORATORY QUESTIONS continued...

5. For **Part C**, make a table summarizing the starting mass of each component and the final mass after the separation. Comment on the possible source of any discrepancies between the two sets of masses.

6. In the space below, write a detailed protocol for the separation of the four-component mixture you performed in **Part C**. Your procedure should be clear enough for another student to exactly follow your steps.

Measurement and Proper Use of Laboratory Glassware

David Roberts

OBJECTIVES

- Correctly use the terms: accuracy, precision and percent error.
- Identify and use the appropriate number of significant figures in measurements and calculations.
- Properly use select laboratory glassware and the analytical balance.
- Calibrate some commonly used glassware.

INTRODUCTION

Your success in science, especially chemistry, will depend on your ability to make measurements that are accurate and consistent. Reliable measurements, beginning with Antoine Laurent Lavoisier, the father of analytical chemistry, have led to our understanding of combustion, chemical composition, and a host of other discoveries. Whether your goal is to make important scientific discoveries, or to just pass this chemistry laboratory, your ability to make careful measurements is an invaluable tool. A big part of making careful scientific measurements involves understanding the limitations of the glassware used to carry out chemical experiments (Figure 1).

Figure 1. Assorted laboratory glassware.

BACKGROUND

Precision and Accuracy

Making quantitative observations or measurements is fundamental to all of science. With any measurement, there are always two factors to consider: accuracy and precision. Accuracy is a measure of how close a measurement is to the actual or theoretical value. Precision is a measure of how close a series of measurements are to one another. These terms are illustrated conceptually in the archer's target (Figure 2). In Figure 2, the target on the left shows an archer who is precise; the arrows are reproducibly clustered around a certain point, but the archer is not accurate because the arrows are not on the bull's eye. If you look at the middle target you will see that the archer was neither accurate nor precise. The arrows are scattered around the target. In the right-hand target, the archer was both accurate and precise. The archer was able to reproduce the same result and able to obtain the desired target.

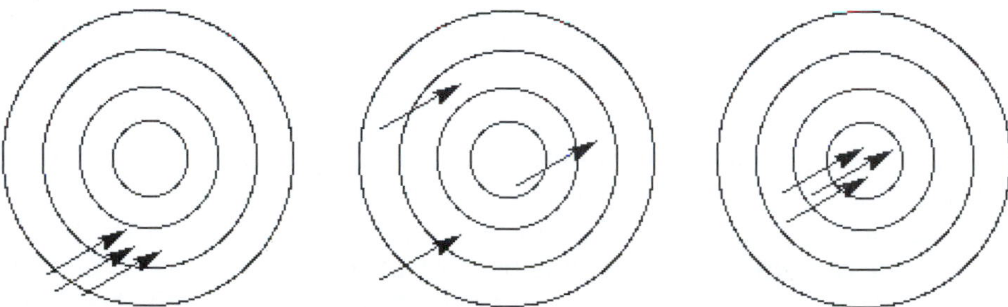

Figure 2. The use of arrows and targets to represent accuracy and precision.

Figure 2 also serves to illustrate the concepts of systematic (or determinate) error and random (or indeterminate) error. The target on the left is an example of systematic error where the archer keeps missing the target in the same fashion; maybe her sight is aligned incorrectly. This type of error can occur in the laboratory if, for example, a balance has been incorrectly calibrated and is adding 0.050 g to each measurement. The target in the middle is an example of random error. In the laboratory, this type of error occurs frequently as a result of sloppy or rushed work on the part of the scientist.

Percent Error

Scientists can evaluate the accuracy of a measurement by comparing the experimental value with the theoretical value. Suppose you measure the mass of a beaker using an analytical balance and the balance reads 283.25 grams. However, the theoretical value for the mass of the beaker is actually 285.00 grams. This difference between the theoretical value and the experimental value is called the error (Equation 1). Error can be either positive or negative depending on whether the experimental value is greater or less than the theoretical value.

$$\text{Error} = \text{Theoretical value} - \text{Experimental value} \qquad \text{Equation 1}$$

For our example, the error is 285.00 g – 283.25 g = +1.75 g. The absolute value of the error is used to calculate the percent error (or relative error) of the measurement (Equation 2).

$$\%\ \text{Error} = \frac{|\ \text{Error}\ |}{\text{Theoretical value}} \times 100 \qquad \text{Equation 2}$$

For our example, $\%\ \text{Error} = \dfrac{1.75\ \text{g}}{285.00\ \text{g}} \times 100 = 0.614\ \%$

Significant Figures and Uncertainty in Measurements

Suppose you wanted to measure your weight using your bathroom scale that is calibrated in 1-pound intervals. On your bathroom scale you could easily read your weight to the nearest pound. In addition, you could also estimate your weight to the nearest tenth of a pound. Suppose you estimate your weight to be between 184 and 185 lbs; let's say 184.5 lbs. The first three digits are known with certainty, but the rightmost digit has been estimated and involves some uncertainty. Scientists often report numbers with plus or minus ranges indicating how uncertain the last digit is. For example, if we thought our weight might be between 184.3 and 184.7 lbs., we could report the weight as 184.5 ± 0.2 lbs. This measurement would have four significant digits. Significant figures include all the digits that are certain, plus the last digit that is estimated. The number of significant figures is related to the certainty of the measurement.

In this experiment you will explore the inherent error in the design of various types of glassware that will affect your ability to obtain accurate measurements. As you will see, it is essential that you understand the limits of the equipment and utilize this information when choosing the appropriate laboratory equipment and glassware.

Laboratory Balance

In the laboratory, we obtain the mass of an object using an electronic analytical balance (Figure 3). Typical laboratory balances can measure masses to within 0.001 g. To guide you in your use of an electronic top-loading balance follow these guidelines:

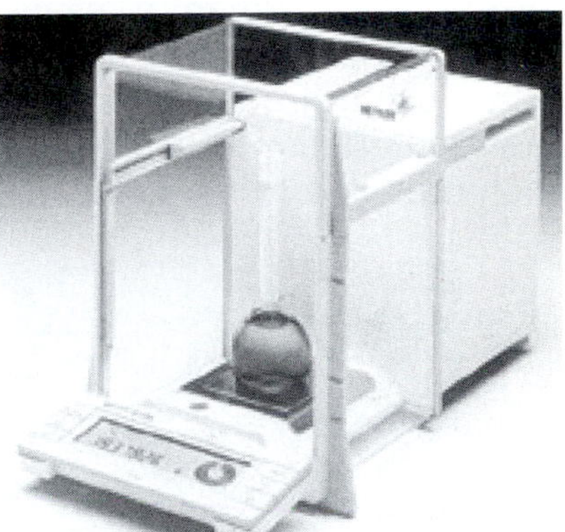

Figure 3. An electronic balance

- The balance should be left on at all times during use.
- Always use the same balance when making repetitive measurements.
- Make certain that all materials are weighed at room temperature.
- Never weigh objects directly on the balance pan. Always use a container or weighing paper on the balance pan.
- Wait for a stable reading before recording the mass.
- When finished, remove your container or weighing paper and clean up any spills.
- You may need to *tare* your container. To do so, press the tare button and wait until the balance shows a stable zero reading. That sets the balance to weigh anything that is placed into the container, but it does not weigh the container itself.

Glassware

When conducting chemical experiments, you will invariably need to use a piece of glassware. The specific glassware you choose depends on the task you are completing. Do you need to obtain as close to 50.00 mL of water as possible? Do you need to pour roughly 3.4 mL of a solution? Do you need to take a solid and make a solution with a carefully known concentration? Or do you just need a container to hold waste?

Beakers, Flasks and Graduated cylinders (Holding Glassware)

Erlenmeyer flasks (Figure 4a) and beakers (Figure 4b) are designed for mixing, transporting, and reacting liquids, but not for accurate volume measurements. The volumes stamped on the sides Erlenmeyer flasks and beakers have considerable error.

Figure 4a. Erlenmeyer flask

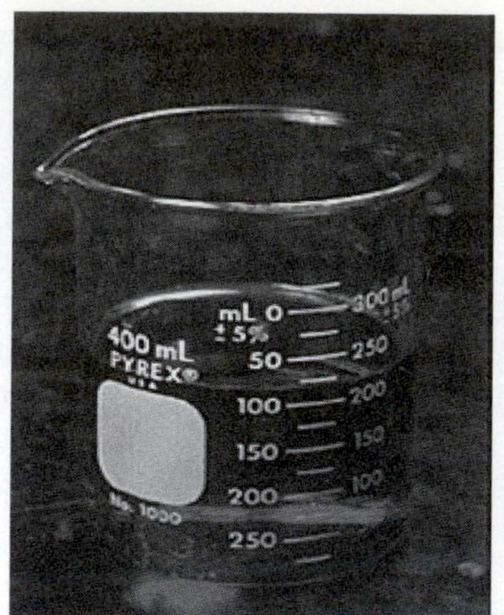

Figure 4b. Beaker

The graduated cylinder (Figure 5a) is a fast and convenient way to dispense an approximate volume of liquid. If you examine the graduated cylinder in Figure 5b, you can see the letters "TD". When a piece of glassware is designated "to deliver" (TD), it has been calibrated to dispense a quantity of liquid knowing that some of the liquid will stay behind. When using graduated cylinders we also have to deal with the tendency of liquids to form a meniscus (Figure 5b). When measuring volumes, it is important to be at eye level with the meniscus and to read the volume marking at the bottom of the meniscus.

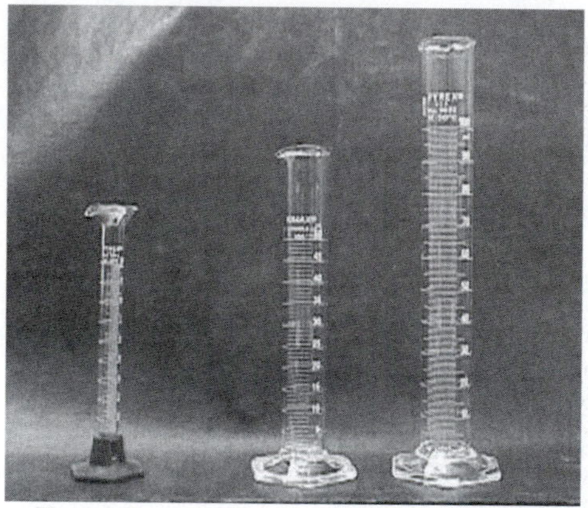

Figure 5a. Various graduated cylinders

Figure 5b. The meniscus

The Volumetric Pipet (Dispensing Glassware)

Volumetric pipets are another form of TD glassware, but unlike graduated cylinders, pipets are used to deliver accurate volumes of liquid. They come in a variety of sizes delivering fractions of a milliliter up to 100 ml or more. The pipet has one of the smallest margins of error for all dispensing glassware. For example, when a 5-ml pipet is used correctly it has been calibrated to deviate only ± 0.01 ml of the target volume. To ensure proper use of the pipet, it is crucial to adhere to the following guidelines:

- Clean the pipet with very dilute soapy water and rinse with de-ionized water.
- From the stock bottle transfer the solution you will pipet into a piece of holding glassware (ex. beaker or flask). Never put a pipet directly into a bottle of stock solution.
- Using a pipet bulb (Figure 6a), carefully fill the pipet with a small amount of the solution. Rotate the pipet so that the inside walls become coated with the solution. Discard the solution in the appropriate waste container.

- Fill the pipet with solution to a point above the calibration mark (be careful not to suck solution up into the bulb) , remove the bulb and quickly cover the opening of the pipet with your index finger (Figure 6b).

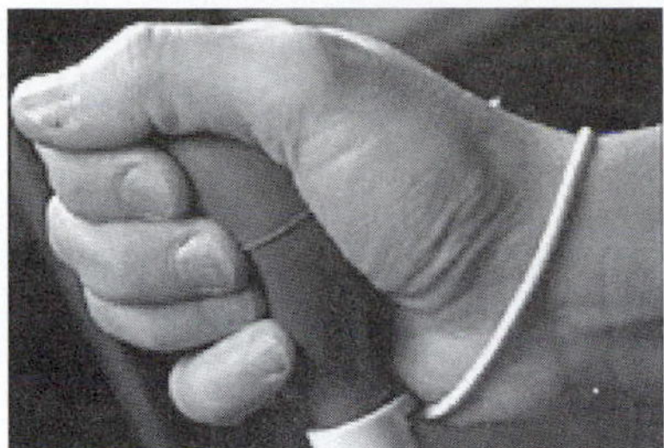

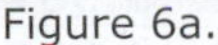

Figure 6a.

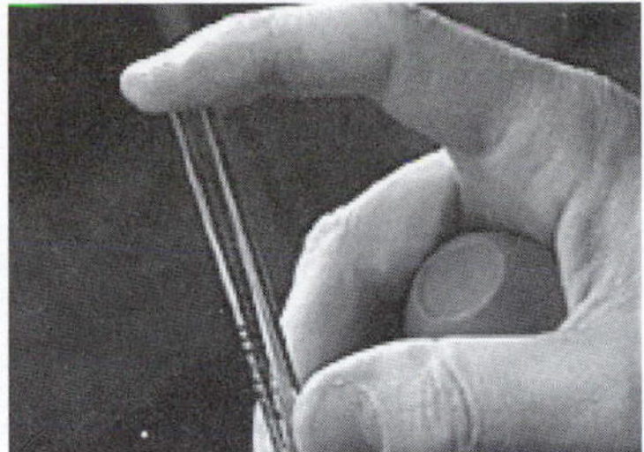

Figure 6b.

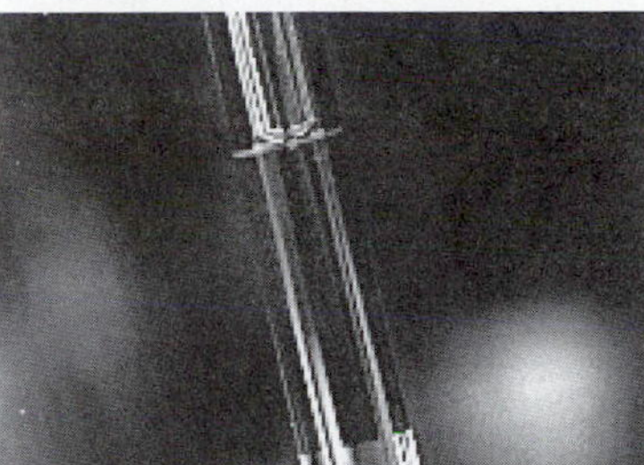

Figure 6c.

- Wipe the tip using a Kimwipe and allow the liquid to drain to the calibration mark. Be sure to read the meniscus at eye level (Figure 6c)
- Touch the tip of the filled pipet to the lower inside wall of receiving glassware and allow the pipet to drain completely (Figure 6d). As with any TD glassware, do not blow out the liquid remaining in the tip of the pipet.

Figure 6d.

Volumetric Flask (Containing Glassware)

Figure 7. A 500-mL volumetric flask.

Volumetric flasks are commonly used to accurately prepare a solution of known concentration. Unlike the glassware previously discussed, the volumetric flask is calibrated to contain (TC) a stated volume of liquid to four significant figures. To prepare solutions using a volumetric flask follow these guidelines:

- Clean and thoroughly rinse the flask.
- Always prepare solutions at room temperature.
- To dilute liquids: deliver a volume of concentrated liquid to the flask. Using a water bottle and a disposable pipet, carefully fill the flask to the calibration mark. Thoroughly mix by capping and inverting.
- To prepare solutions from solids: Completely dissolve the solid in a minimum amount of solvent. Using a water bottle and a disposable pipet, carefully fill the flask to the calibration mark. Thoroughly mix by capping and inverting.

Calculating Error and Percent Error of Select Glassware

In this laboratory exercise, the percent error of a piece of glassware will be assessed using the relationship between the mass and density of water. The error in any type of glassware is the difference between the actual volume delivered (based on the relationship between mass and density) and the theoretical volume (actual volume marking indicated on the glassware). Table 1 lists the density of water at various temperatures, which will aid you in completing this experiment.

Table 1. Density of water as a function of temperature

Temperature (°C)	Density of water (g/mL)	Temperature (°C)	Density of water (g/mL)
15	0.9991026	23	0.9975415
16	0.9989460	24	0.9972995
17	0.9987779	25	0.9970479
18	0.9985986	26	0.9967867
19	0.9984082	27	0.9965162
20	0.9982071	28	0.9962365
21	0.9979955	29	0.9959478
22	0.9977735	30	0.9956502

OVERVIEW

In this experiment you will practice proper laboratory techniques. You will then determine the accuracy of a 100-mL graduated cylinder, a 50-mL pipet and a 250-mL beakers by comparing your results to the rest of the class. Finally you will use a volumetric flask to make a solution of known concentration. The solution will then be diluted.

PROCEDURE

PART A. Performance Objectives

Materials used per pair	
50-mL Graduated cylinder 50-mL Pipet and bulb 100-mL Volumetric flask	250-mL Beaker Top loading balance Disposable pipet

Part A of this experiment is an opportunity for you to learn the proper procedure to "determine the mass of 50 mL of water transferred from a graduated cylinder to a beaker" and to "accurately deliver 50 mL of water to a volumetric flask". To obtain the most benefit from this exercise, review the information in the BACKGROUND pertaining to the handling of laboratory glassware. Then, practice performing the steps outlined in the following two "Tasks". Once you feel comfortable that you can perform the "Tasks" without the aid of your laboratory manual, give your manual to your laboratory partner and have them evaluate your work. For each step that you correctly perform, your partner will mark a √ in the corresponding column. After completing the task, evaluate your partner's performance.

Task A: Determine the mass of 50 mL of water transferred from a graduated cylinder to a beaker.

Performance objective	**OK?**
Clean graduated cylinder with diluted soapy water and rinse thoroughly.	
Carefully pour water into the cylinder until close to the desired volume.	
Reach final volume by using a disposable pipet to add water drop wise.	
Student reads the bottom of the meniscus at eye level to ensure accuracy.	
Place a clean, dry beaker on the balance pan.	
Wait for a stable reading and record the weight of the beaker.	
Leave the balance neat and clean.	
Transfer the water from the graduated cylinder into the beaker.	
Place the beaker with water onto the same balance used previously.	
Wait for a stable reading and record the weight.	
Leave the balance neat and clean.	

Task B: Accurately deliver 50 mL of water to a volumetric flask.

Performance objective	**OK?**
Clean and dry a beaker.	
Pour a volume of water, in excess of the desired amount (in this case, 50 mL) into the beaker.	
Clean the pipet with dilute, soapy water and rinse thoroughly.	
Rinse the pipet with the solution of interest (in this case, water).	
Draw the solution up past the calibration mark without sucking the solution into the bulb.	
Remove the bulb and quickly cap the pipet with an index finger.	
While at eye level with the pipet, slowly lower the solution to the calibration mark.	
Make sure the bottom of the meniscus is at the calibration mark.	
Touch the tip of the filled pipet with Kimwipe to remove excess solution.	
Lower the pipet inside the receiving glassware (in this case, a volumetric flask) and allow the pipet to completely drain.	
Do not blow out the last drop of solution on any "to deliver" (TD) glassware.	

PART B. Glassware Calibration

Materials Used per Individual	
100-mL Graduated cylinder 50-mL Pipet and bulb 100-mL and 250-mL Beakers	Analytical balance Thermometer

Weigh a clean, dry 100-mL beaker. Measure 50 mL of water using a 100-mL graduated cylinder. Carefully transfer the water from the graduated cylinder to the empty 100-mL beaker. Re-weigh the beaker. Record the temperature of the water in the beaker. Repeat this process two more times for the graduated cylinder.

Repeat the above process (three trials) using, first, a 50-mL pipet and then a 250-mL beaker in place of the graduated cylinder. Record your average experimental values for the volumes of water dispensed by each piece of glassware on the board with the class data. Be sure to record all of the class values in your notebook for later analysis.

PART C. Solution Preparation Using a Volumetric Flask

Chemicals Used	Materials Used
$Cu(SO_4){\bullet}5H_2O$ (s)	5-mL Pipet and bulb 250-mL Beaker (1) 100-mL Volumetric flask (2)

Weigh 0.500 g of $Cu(SO_4){\bullet}5H_2O$ (s) on a piece of weighing paper. Transfer the solid to a 100-mL volumetric flask. Make Solution #1 by dissolving the solid and then diluting to the mark according to the steps described in the BACKGROUND for using a volumetric flask.

Next, prepare Solution #2 by pipeting 5.00 mL of the Solution #1 into a second 100-mL volumetric flask and diluting to the mark. Record your observations comparing Solution #1 to Solution #2.

Measurement and Proper Use of Laboratory Glassware

Name:	Lab Instructor:
Date:	Lab Section:

PRE-LABORATORY EXERCISES

1. Distinguish between the terms "mass" and "weight".

2. Using your textbook as a guide, summarize the rules for determining if a zero in a number is a significant figure. Give an example where a zero is significant and an example where it is not significant. For your two examples, also indicate how many significant figures are in each number.

3. Summarize the rules that apply to addition/subtraction operations and multiplication/division operations for significant figures.

4. Perform the following calculations. Include correct units and indicate the number of significant figures in each answer
   a) 1.456 m x 205 m

   b) 1.456 m + 205 m

OVER →

PRE-LABORATORY EXERCISES continued...

5. A student conducts three trials to determine the concentration of an unknown solution. She reports the following values: 0.210 M, 0.198 M and 0.203 M. The actual concentration is 0.135 M.
   a) What is her average reported value?

   b) What is her error?

   c) What is her percent error?

   d) Was her work accurate? Was it precise? Explain.

   e) Did her work show systematic error? Did it show random error? Explain.

6. A small, empty container is found to weigh 2.356 g. The container is then filled with water (23°C). The full container is found to weigh 28.624 g. Using the information in Table 1 in the BACKGROUND, determine the volume of water (mL) in the beaker.

Measurement and Proper Use of Laboratory Glassware

Name:	Lab Instructor:
Date:	Lab Section:

RESULTS and POST-LABORATORY QUESTIONS

PART A. Performance Objectives

Are there any steps that you forgot to perform when your laboratory partner evaluated your ability to carry out Tasks A and B? If so, which steps?

PART B. Glassware Calibration

Attach copies of your data tables (one for each type of glassware tested). Your tables must include the following information for each of the three trials: Mass of empty glassware, mass of glassware and water, mass of water, temperature of water, corresponding density of water, volume of water, your experimental average, your error (compared to the average) and your percent error (again for the average and assuming a theoretical volume of 50.00 mL). Be careful to report all data with proper units and with the correct number of significant figures.

Beginning with the mass of the empty graduated cylinder, show all of your work for calculating the volume of water dispensed in Trial #1 with the graduated cylinder.

Beginning with the volumes of water dispensed by the graduated cylinder in the three trials, show all your work for calculating the average, the error (using the average) and the percent error (using the average experimental value) for the graduated cylinder.

What is the class average for the experimental volume of water dispensed by each piece of glassware. For each type of glassware, what is the class error (using the class average) and the class percent error. Based on these experimental results, which is the most accurate piece of glassware? Explain.

OVER →

RESULTS and POST-LABORATORY QUESTIONS continued...

PART B. Glassware Calibration continued...

For each piece of glassware indicate the range of class values (highest and lowest) for the experimental volume of water. Based on these experimental results, which is the most precise piece of glassware? Explain.

Which of the following pieces of glassware (volumetric flask, beaker, pipet or graduated cylinder) would be best suitable for the following tasks? Briefly explain each answer.

a) Use as a waste container.

b) To dilute a solution to a concentration with 4-significant figures.

c) Accurately deliver 25.00 mL of solution.

d) Quickly deliver approximately 25 mL of solution.

PART C. Solution Preparation using a Volumetric Flask

What is the concentration of Solution #1? Report your answer with the correct number of significant figures in terms of molarity (moles of $Cu(SO_4)\bullet5H_2O$/liter of solution).

What did you observe by comparing Solution #1 to Solution #2? Does this make sense in terms of the relative concentrations? Explain.

Chemical Nomenclature, Part I: Naming Ionic Compounds

Holly Garrison and Kristen Spotz

OBJECTIVES

- Distinguish between molecular and ionic compounds.
- Know the charges on common polyatomic ions.
- Write appropriate chemical formulas for ionic compounds.
- Name ionic compounds.

INTRODUCTION

Beginning in the 1st century AD, alchemists conducted numerous chemical reactions. Soon, alchemists developed a novel vocabulary for describing the chemicals they experimented with. The metals, for example, were named after astrological bodies (ex. gold was called the sun and silver was called the moon). Alchemists also named elements or compounds after an observable property such as appearance or smell (ex. copper acetate has a greenish hue so it was called Spanish green). In addition, many elements or compounds were named after their discoverer or the place where discovered (ex. Epsom salt). Soon, this method of naming became vast and confusing with chemical names varying from place to place, making communication almost impossible. To deal with the over 13 million known chemical substances, scientists have devised a universal system for naming compounds.

BACKGROUND

Ionic Compounds

Although a few elements occur uncombined in nature (most notably the noble gases), the majority of elements form compounds. Whether the compound is an ionic compound or a molecular compound can usually be determined by noting the types of elements making up the compound. When different non-metallic elements combine, a molecular compound is formed. If, however, metallic elements combine with non-metallic elements, an ionic compound is typically formed. In this laboratory experiment, we will focus on ionic compounds. In ionic compounds, the metal transfers one or more of its electrons to the non-metal resulting in ions that are electrostatically attracted to each other to form a neutral compound. An important exception you should know to the generalization that all cations are metals, is the ammonium ion, NH_4^+.

Ionic compounds consist of an array of cations and anions extending in all three dimensions (Figure 1). In addition, ionic compounds are neutral, meaning they possess no net charge. For this to occur, the ionic compound must possess equal numbers of positive and negative charges not necessarily equal numbers of cations and anions. If a cation has a plus one charge, and an anion has a negative one charge, they can react in a 1:1 ratio to form a neutral compound. Is is important that you learn the charges on the common monotomic ions (Table 1). For example, magnesium oxide, MgO, contains Mg^{2+} ions and O^{2-} ions in a 1:1 ratio. It is more complicated when elements combine whose ions have different charges. For example, the ionic compound formed from magnesium and chlorine contains Mg^{2+} ions and Cl^- ions. To be neutral, one Mg^{2+} ion is combined with two Cl^- ions, resulting in the formula $MgCl_2$.

Figure 1. The three-dimensional arrangement of sodium cations and chloride anions in NaCl.

Period	1A (1)	2A (2)	3B (3)	4B (4)	5B (5)	6B (6)	7B (7)	8B (8)	8B (9)	8B (10)	1B (11)	2B (12)	3A (13)	4A (14)	5A (15)	6A (16)	7A (17)	8A (18)
1	H^+																H^-	
2	Li^+														N^{3-}	O^{2-}	F^-	
3	Na^+	Mg^{2+}											Al^{3+}			S^{2-}	Cl^-	
4	K^+	Ca^{2+}				Cr^{2+} Cr^{3+}	Mn^{2+}	Fe^{2+} Fe^{3+}	Co^{2+} Co^{3+}		Cu^+ Cu^{2+}	Zn^{2+}					Br^-	
5	Rb^+	Sr^{2+}									Ag^+	Cd^{2+}		Sn^{2+} Sn^{4+}			I^-	
6	Cs^+	Ba^{2+}										Hg_2^{2+} Hg^{2+}		Pb^{2+} Pb^{4+}				
7																		

Table 1. Charges on the common monatomic ions. Note that some of the ions have two charges listed. We will look further at these ions in the next section.

Naming Binary Ionic Compounds

Both magnesium oxide and magnesium chloride are examples of binary ionic compounds. Binary ionic compounds are formed from monatomic ions, which are either anions or cations derived from a single atom. Once you have learned the charges of the common monatomic ions, you can write their chemical formulas and name them. From Table 1, we see that the main-group elements (1A, 2A, 3A, 5A, 6A, 7A) take on a single particular charge when they form ions. Many transition elements, however, form two different monatomic ions (ex. Cu^+ and Cu^{2+}). The distinction between the different charges is noted in the name of a compound. Binary ionic compounds are named in the following manner:

- The complete name of the cation is the first part of the chemical name and is the same name as the metal.
- The name of the anion is the second part of the chemical name and takes the root of the non-metal plus the suffix "ide".
- For metals that have more than one common ionic charge, Roman numerals are used in the name to distinguish between the different possible ions. Do not use Roman numerals for elements that only form one common monatomic ion.

Example 1: We have already seen that Mg^{2+} ions and Cl^- ions combine in a 1:2 ratio to form $MgCl_2$. This compound is named by first identifying the metal cation, **magnesium**, and then identifying the non-metal, chlorine. The suffix "ide" is added to the root of the non-metal to produce **chloride**. The name of the metal comes first, followed by the changed name of the anion. The complete chemical name for $MgCl_2$ is **magnesium chloride**.

Example 2: The ammonium ion, NH_4^+, and the chloride ion, Cl-, combine to form NH_4Cl, **ammonium chloride**. Again, the cation is named first and the root of the anion with the attached suffix, "ide", is named second.

Example 3: To avoid the ambiguity in naming ionic compounds whose metal ions have more than one possible charge, a Roman numeral is placed inside parentheses and immediately follows the name of the metal. For example, Fe^{2+} and Fe^{3+} are name iron (II) and iron (III) respectively. The compound formed when Fe^{3+} combines with O^{2-} in a 2:3 ratio, Fe_2O_3, is called **iron(III) oxide**. Your textbook probably describes a second older method for designating the names of ions with more than one charge by using the suffixes –ous and –ic. In this laboratory, we will only use the Roman numeral method.

Naming Polyatomic Ions

Many ionic compounds contain one or more polyatomic ions (Figure 2). Polyatomic ions consist of two or more atoms held together by covalent bonds and possessing either a net positive or negative charge. Polyatomic ions react as a single unit, staying together during interactions with other ions. You have already seen the example of the ammonium cation. Table 2 lists the names of some more polyatomic ions that you will encounter in your study of chemistry. The highlighted ions are the most common.

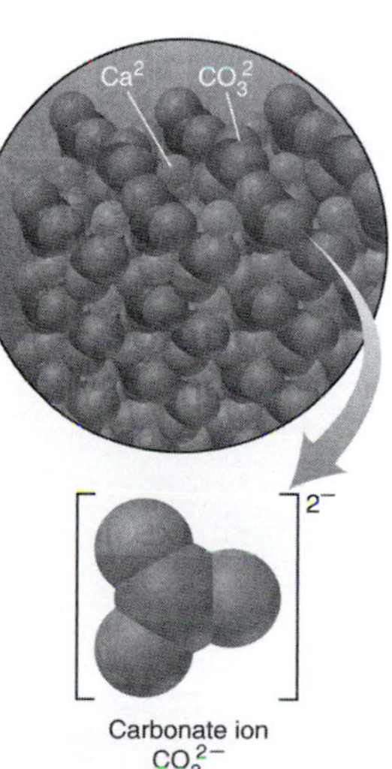

Figure 2. Carbonate, CO_3^{2-}, is a polyatomic ion. The carbonate ion consists of four covalently bound atoms acting as a single unit carrying an overall -2 charge.

Table 2. The names of some common polyatomic ions.

Formula	Name	Formula	Name
$C_2H_3O_2^-$	**acetate**	**CO_3^{2-}**	**carbonate**
CN^-	cyanide	**HCO_3^-**	**bicarbonate**
OH^-	**hydroxide**	CrO_4^{2-}	chromate
ClO^-	hypochlorite	**$Cr_2O_7^{2-}$**	**dichromate**
ClO_2^-	chlorite	**PO_4^{3-}**	**phosphate**
ClO_3^-	**chlorate**	HPO_4^{2-}	hydrogen phosphate
ClO_4^-	**perchlorate**	$H_2PO_4^-$	dihydrogen phosphate
NO_2^-	nitrite	SO_3^{2-}	sulfite
NO_3^-	**nitrate**	**SO_4^{2-}**	**sulfate**
MnO_4^-	**permanganate**	HSO_4^-	hydrogen sulfate

Most of the polyatomic ions in Table 2 have an non-metal element bonded to one or more oxygen atoms. These polyatomic ions are called oxoanions. In several cases, there are groupings of two to four oxoanions that differ only in the number of oxygen atoms.

The guidelines for naming compound with polyatomic ions are the same as for naming binary ionic compounds. The compound formed when sodium ions, Na^+, combine with nitrate ions, NO_3^-, is called **sodium nitrate**, $NaNO_3$. Copper(II) ions, Cu^{2+}, combine with chlorate ions, ClO_3^-, in a 1:2 ratio. The resulting compound, $Cu(ClO_3)_2$, is called copper(II) chlorate. Notice, parenthesis are placed around polyatomic ions when two or more of the same ion is present in a particular compound.

OVERVIEW

In this experiment, you will name ionic compounds and write their chemical formulas. You will begin by playing "BINGO" to learn the names of the polyatomic ions. You will play "Jigsaw" to help you see how the various ions combine to form neutral ionic compounds. Finally, you will play "The Ultimate Naming Game" allowing you to show your complete mastery in naming ionic compounds.

PROCEDURE

Part A. BINGO

Materials Used
"BINGO" game board (found at the end of the post-laboratory questions) Place Markers

Working individually, select any 64 of the following names and formulas:

Names: iron(III), cadmium, potassium, sodium, copper(I), chromium(II), ammonium, aluminum, zinc, barium, lithium, cobalt(II), calcium, silver, mercury(II), lead(IV), sulfide, oxide, nitride, hydroxide, dichromate, nitrite, nitrate, cyanide, perchlorate, carbonate, phosphate, acetate, hydride, fluoride, chloride, iodide.

Formulas: Sn^{2+}, Cu^{2+}, Cr^{3+}, Na^{+}, K^{+}, Ca^{2+}, Fe^{2+}, Pb^{2+}, Ca^{2+}, Li^{+}, Ag^{+}, Zn^{2+}, Cd^{2+}, Al^{3+}, NH_4^{+}, Mn^{2+}, NO_2^{-}, $C_2H_3O_2^{-}$, OH^{-}, CN^{-}, S^{2-}, H^{-}, F^{-}, Br^{-}, NO_3^{-}, I^{-}, N^{3-}, O^{2-}, MnO_4^{-}, HCO_3^{-}, PO_4^{3-}, $Cl0_4^{-}$.

Remove the BINGO game board found at the end of the post-laboratory questions. Write one of the names or formulas you chose from the above list in each of the game board squares (except the FREE square).

To play the game: If your laboratory instructor calls (or writes on the board) a name of an ion, find the corresponding formula on your game board and mark the square. For example if they say "nitrate", you try to find "NO_3^{-}" on your game board. If your instructor calls (or writes on the board) a formula, find the corresponding name on your game board and mark the square. The first person to fill in five squares vertically, horizontally or diagonally yells "BINGO". Use the second copy of the BINGO gam board if your instructor plays the game a second time. During the game, copy down in your laboratory manual the list of formulas and names called regardless of whether or not they are on your BINGO card.

Part B. Jigsaw

Materials Used
Jig-Saw puzzle pieces (found at the end of the post-laboratory questions) scissors

Form a group of 3 - 4 students. Remove the Jig-Saw puzzle pieces found at the end of the post-laboratory questions. Have each student cut out all of their puzzle pieces. Combine the puzzle pieces from all of the students in the group.

Taking turns, form all sixteen possible compounds using the puzzle pieces. Note that each puzzle piece has a different number of peaks (cations) and valleys (anions) depending on the ion charge. By using the correct number of each of the pieces, you will see in what ratio the ions combine to form the final, neutral compound. As you build each combination, record in your laboratory notebook, the symbol and name of each ion, the ratio in which they combine and the final formula of the compound.

For example: Na^{+} (sodium) and SO_4^{2-} (sulfate) combine in a 2:1 ratio to form Na_2SO_4.

Part C. The Ultimate Naming Game

Materials Used
"The Ultimate Naming Game" game boards (found at the end of the post-laboratory questions)
Die
Place markers

Form a group of 3 – 4 students. Take turns rolling the die. Each student rolls the die three times. The first roll indicates which set of game boards that student will play with for that turn. The first page of game boards should be used when an EVEN number is rolled and the second page of game boards is used when an ODD number is rolled. The second roll of the die indicates how many spaces to move the place marker on the first game board (CATIONS). The die is then rolled a third time indicating how many spaces to move the place marker on the second board (ANIONS).

When playing with the EVEN number boards, *each* student should write down the name of the CATION and ANION that was rolled and the name and formula for the resulting complex that is formed. In the case of cations that form more than one oxidation state, write all possible names and formulas. For example, if "Copper" and "Nitrate" are rolled, write down the names and formulas of the compounds: Copper (I) nitrate, $CuNO_3$ and Copper (II) nitrate, $Cu(NO_3)_2$.

When playing with the ODD number boards, *each* student should write down the formula of the CATIONS and ANIONS rolled and the name and formula of the resulting compound. When naming compounds, remember to decide whether roman numerals are needed. Answers for the groups should then be compared. Continue to play until your instructor tells you the game is over (roughly 20 – 30 minutes). Your instructor will also provide you with time to ask any questions you had as a group.

Chemical Nomenclature, Part I: Naming Ionic Compounds

Name:	Lab Instructor:
Date:	Lab Section:

PRE-LABORATORY EXERCISES

1. Define the terms "metal" and "non-metal". Where are the metals and the non-metals located on the periodic table?

2. For each of the following compounds, identify each element as either a metal or a non-metal and the compound as either ionic or covalent.

   a) N_2O_5

   c) MgO

   b) CO_2

   d) Al_2O_3

3. What is the chemical formula and name for the compound occurring when Ni^{2+} ions combine with PO_4^{3-} ions?

4. Refering to your textbook, summarize the rules for naming oxoanions. Work through an example by naming the chlorine family of oxoanions.

Chemical Nomenclature, Part I: Naming Ionic Compounds

Name:	Lab Instructor:
Date:	Lab Section:

RESULTS and POST-LAB QUESTIONS

Part A. BINGO
Attach a copy of all of the formulas and names called during the BINGO game. For each formula called, write the name and for each name called, write the formula.

Part B. Jigsaw
Attach a copy of the symbol and name of each ion, the ratio in which they combine and the final formula of the sixteen different compounds.

Part C. The Ultimate Naming Game
Attach a copy of the names and formulas of all the compounds that your group rolled during the game. For compounds formed from metals that form more that one oxidation state, be sure to give the names and formulas of both of the possible compounds.

Game boards for Part A: BINGO

		FREE		

		FREE		

Puzzle pieces for Part B: Jigsaw

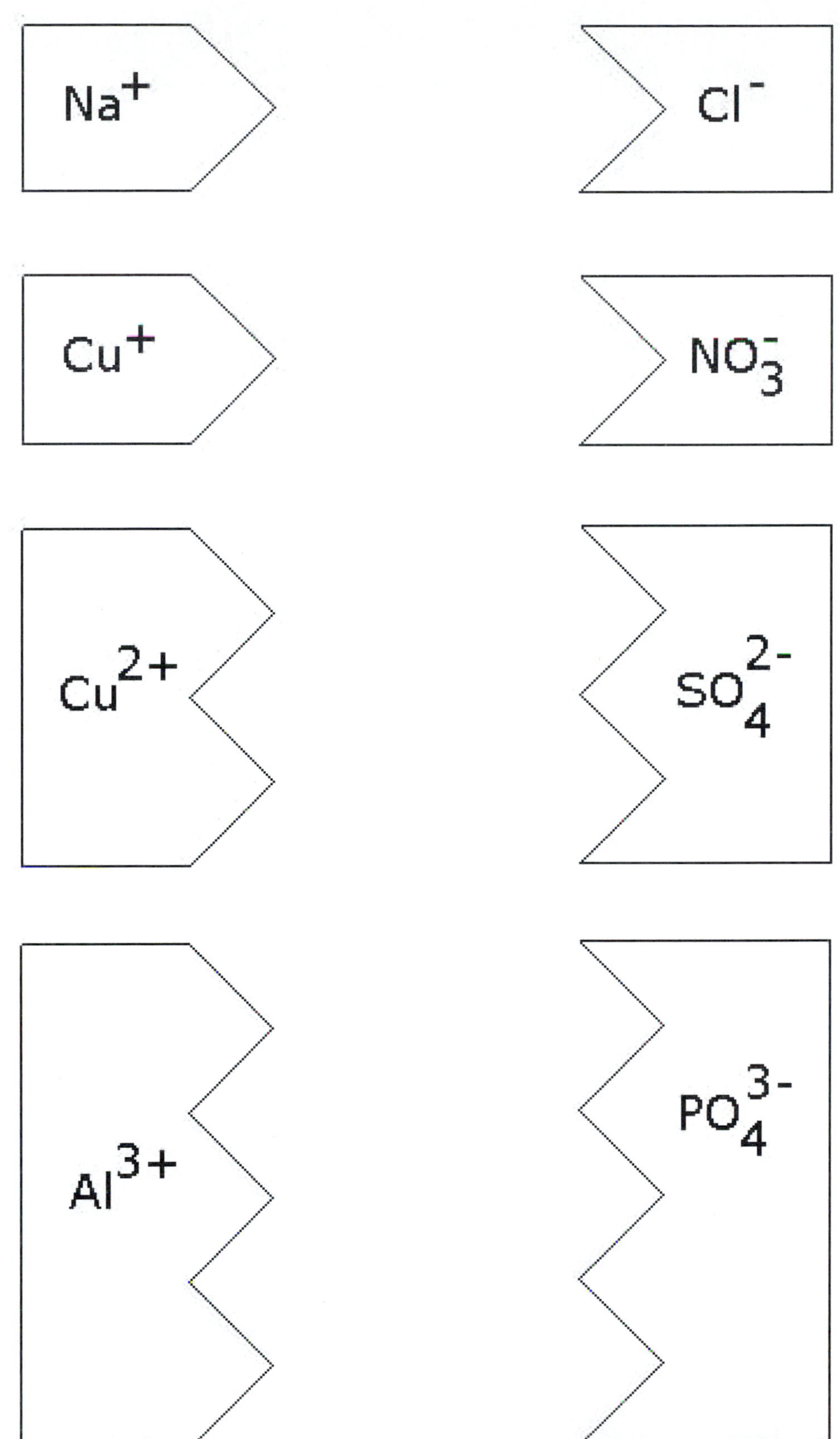

Game boards for Part C: The Ultimate Naming Game
Use these boards for an **EVEN** roll of the die.

CATIONS

START → Cobalt	Lithium	Barium	Zinc
Calcium			Aluminum
Silver			Ammonium
Mercury			Chromium
Lead			Copper
Iron	Cadmium	Potassium	Sodium

ANIONS

START → Fluoride	Chloride	Iodide	Sulfide
Bicarbonate			Oxide
Acetate			Nitride
Phosphate			Hydroxide
Carbonate			Dichromate
Perchlorate	Cyanide	Nitrate	Nitrite

Game boards for Part C: The Ultimate Naming Game

Use these boards for an **ODD** roll of the die.

CATIONS

START → Ca^{2+}	Li^{+}	Ag^{+}	Zn^{2+}
Pb^{2+}			Cd^{2+}
Mg^{2+}			Al^{3+}
Ba^{2+}			NH_4^{+}
K^{+}			Mn^{2+}
Na^{+}	Cr^{3+}	Cu^{2+}	Sn^{2+}

ANIONS

START → Cl^{-}	F^{-}	Br^{-}	NO_3^{-}
S^{2-}			I^{-}
CN^{-}			N^{3-}
OH^{-}			O^{2-}
$C_2H_3O_2^{-}$			MnO_4^{-}
NO_2^{-}	ClO_4^{-}	PO_4^{3-}	HCO_3^{-}

Understanding the Basics: The Mole and Counting Atoms

Brian Polk

OBJECTIVE

- Explain why the mole is a useful unit in chemistry.
- Be able to count things using mass.
- Convert between grams $\leftrightarrow$ moles $\leftrightarrow$ atoms/molecules.

INTRODUCTION

Chemists routinely work with extremely small things, such as atoms and molecules, that are, for practical purposes, too small to count by conventional methods. The mole is a convenient unit for counting very small things. By the end of this laboratory exercise, you will have an appreciation for the fact that a mole is really no different than other counting units, such as a dozen, a ream or a pair. The key to really understanding the mole however, is in realizing the huge order of magnitude involved.

BACKGROUND

Atoms are made up of protons, neutrons and electrons. The atomic number tells us the number of protons in an atom, and the mass number indicates the number of protons plus neutrons. Taking into account the various isotopes of an element, the average atomic mass can be determined. Previously in your studies, atomic mass has been defined in terms of atomic mass units (amu). For example, an average atom of iron weighs 55.845 amu, as seen on the periodic table (Figure 1). These units, however, are not very practical for chemists because they are far too small to measure directly (1 amu = 1.66×10^{-24} g). The solution to this problem is to scale-up this measurement into a convenient unit. A unit is needed that can be easily converted to a measurable quantity like grams. As you know, grams can easily be weighed with an analytical balance. The unit of choice for this scale-up is the mole.

26 **Fe** 55.845

Figure 1

Methods of Counting

When we talk about quantifying something, two main methods surface. For example, you can buy flour by the pound. Flour is **quantified by mass**. For other substances, it is easier to **quantify by number**. For example, a ream of paper always contains 500 sheets (though the mass will vary depending on the size and thickness of each sheet). In chemistry, however, we have the best of both worlds. By using the mole, we can quantify a substance by both mass and number.

To understand what a mole is, let's first think about a scale-up unit that you are already familiar with, such as a dozen. Why are donuts packaged in dozens? Twelve seems to be a number that works well for selling and packaging donuts. Of course, you can buy fewer or more, but the dozen make a nice fundamental unit. For the chemist, one dozen is still far too small. A dozen atoms of carbon, for example, would only weigh 2.39×10^{-22} g. Even one billion carbon atoms is too small to easily weigh. If we want to work with things as unbelievably small as atoms, then we'll need a scale-up number that is unimaginably large.

The most recent experimental value of a mole puts it at $6.022136736 \times 10^{23}$ things (just like a dozen is twelve things). For our purposes, though, we don't need all those significant figures, and 6.022×10^{23} usually works just fine. This number is referred to as Avogadro's number, N_A, in honor of the Italian scientist who came up with the concept. It is interesting to note that in Avogadro's lifetime, he had no way of calculating the value of his number.

You can have a mole of anything, and although it certainly is a big number, it is still just a number (Figure 2). A mole of cars means you have 6.022×10^{23} cars. A mole of water molecules is 6.022×10^{23} water molecules.

Figure 2. One mole of some common objects: chalk, oxygen, copper and water (from left to right)

If you were able to place one carbon atom at a time onto an analytical balance, by the time you reached 6.022×10^{23} atoms, the balance would read 12.01 g (look at carbon on the periodic table, carbon weighs 12.01 amu), and you would be very tired and old! Likewise, you can weigh out 12.01 g of carbon atoms and know that you have 6.022×10^{23} carbon atoms (1 mole of carbon atoms) without having to count them! The term that makes these interconversions possible is called the molar mass.

The molar mass is of fundamental importance to chemists because it gives the mass of one mole of a specific object and, therefore, allows us to convert between grams (**mass**) and moles (**number**). You will practice these conversions in the pre-lab exercises.

1 mole carbon = 12.011 g carbon = 6.022×10^{23} carbon atoms

1 mole iron = 55.845 g iron = 6.022×10^{23} iron atoms

OVERVIEW

In this experiment we will explore the difference between quantifying by mass and by number by weighing pennies, nickels, beans and rice. We will then measure mole quantities of a few common substances like water and iron, and finally figure out how many carbon atoms are in a packet of sugar.

PROCEDURE

Chemicals Used	Material Used
Iron cube Water Packet of sugar	Balance Bag of nickels, bag of pennies Small beaker Bag of beans and rice

Part A. "Element #1": the Penny

Determine and record the following information: the weight of a single penny, the weight of the "known" bag of pennies (Bag A), the number of pennies in the "known" bag (by counting them). The weight of the "unknown" bag of pennies (Bag B). Be sure to subtract the mass of the empty bags (written on the bags) from the total masses.

Part B. "Element #2": the Nickel

Determine and record the following information: the weight of a single nickel.

Part C. "Compound #1": the Penny-Nickel

We can now combine our penny "element" with our nickel "element" to make various "compounds". Figure 3 shows a single "molecule" composed of two pennies and one nickel.

Figure 3. Pennies and nickels forming a compound in a 2:1 ratio

Determine and record the weight of a bag containing "molecules" composed of 2 pennies for every 1 nickel (Bag C). Do not remove the contents from the bag, but be sure to account for the mass of the bag.

Part D. "Compound #2": the Bean-Rice

Determine the mass of 10 grains of rice and 10 beans. Determine and record the weight of a bag containing a "molecules" composed of beans and rice (Bag D). Record the ratio of beans-to-rice in a single molecule (written on the bag). Again, do not remove the contents, but be sure to account for the mass of the bag.

Part E. Converting from Mass to Moles for Some Common Objects

Record the weight of each of the following: a small iron cube, 50 mL of water and the contents of a packet of sugar.

Understanding the Basics: The Mole and Counting Atoms

Name:	Lab Instructor:
Date:	Lab Section:

PRE-LABORATORY EXERCISES

1. Define the underlined words in the BACKGROUND section.

2. Write out Avogadro's number using non-scientific notation.

3. What is the mass (in grams) of one mole of gold atoms? Assuming that the price of gold is \$11,000.00/kg, what is the approximate \$ value of one mole of gold atoms? Which would you rather have, one mole of pennies (6.02×10^{23} pennies) or one mole of gold atoms (6.02×10^{23} atoms)?

4. Which is greater in terms of mass, a mole of sodium atoms or a mole of sulfur atoms? Which of the two is greater in terms of number of atoms?

5. Could donuts be quantified in terms of mass instead of number? Explain.

6. Invent a term to describe a certain number of something you use that doesn't already have a convenient means of quantification.

Understanding the Basics: The Mole and Counting Atoms

Name:	Lab Instructor:
Date:	Lab Section:

RESULTS and POST-LAB QUESTIONS

Part A. "Element #1": the Penny

1. Mass of single penny:
2. Mass of the pennies in the "known" bag (Bag A):
3. Number of pennies counted in the "known" bag:
4. Using the mass of a single penny, calculate the number of pennies in the "known" bag of pennies. How does this number compare to the counted number of pennies?
5. Mass of the pennies in the "unknown" bag (Bag B):
6. Without counting the number of pennies in the "unknown" bag (Bag B), calculate the number of pennies in the "unknown" bag.

Part B. "Element #2": the Nickel

1. Mass of single nickel:
2. How much (in kg) would a bag containing $15,000 worth of nickels weigh?

Part C. "Compound #1": the Penny-Nickel

1. Mass of a single penny-nickel "molecule":
2. Mass of the penny-nickel "molecules" in the bag (Bag C).
3. Without counting the contents of the bag, calculate the number of penny-nickel "molecules" in the bag (Bag C).
4. You are a chemist and you have made a new "compound" composed of just pennies and nickels (in a different ratio than the 2:1 ratio seen above). A 13.7 kg sample of this new "compound" is separated into its component "elements" with the following results: 3.84 kg of pennies and 9.90 kg of nickels. How many pennies are present in the 13.7 kg sample? How many nickels are present? What is the ratio of pennies to nickels present in the new "compound"?

OVER →

RESULTS and POST-LAB QUESTIONS continued...

Part D. "Compound #2": the Bean-Rice

1. Mass of 10 beans:
2. Mass of 10 grains of rice:
3. Using the bean-to-rice ratio from the bag, calculate the average mass of one bean-rice "molecule".
4. Mass of the bean-rice "molecules" in the bag (Bag D).
5. Without counting the contents of the bag, calculate the number of bean-rice "molecules" in the bag (Bag D).

Part E. Converting from Mass to Moles for Some Common Objects

1. How much did the iron cube weigh? Calculate the number of moles of iron and the number of iron atoms in the cube.
2. How much did the 50 mL of water weigh? How many molecules are there? How many atoms of hydrogen are there? Atoms of oxygen? About how many water molecules do you drink per day? How many mL of water should you weigh out to have one mole of water?
3. What was the mass of the contents of the sugar packet? Calculate the number of moles of sugar in the packet (assume the sugar is pure sucrose). How many molecules of sugar are in the packe? Calculate the numbers of atoms of carbon, oxygen, and hydrogen in the packet.

Limiting Reactants: How Much $BaSO_4$ Can We Make?

Holly Garrison

OBJECTIVES

- Be able to write and balance chemical equations.
- Perform calculations involving limiting reactants, theoretical yield, actual yield and percent yield.

INTRODUCTION

You have been asked to make ham and cheese sandwiches (Figure 1) for a group of children. Each sandwich requires two pieces of bread, one slice of ham, two slices of cheese, and one piece of lettuce. You have three loaves of bread each containing twenty-four slices, four packages of cheese containing sixteen slices each, six packages of ham each containing eight slices, and three heads of lettuce containing fourteen pieces. Which ingredient limits the number of sandwiches you can make? Which ingredients are in excess?

2 slices of bread + 1 slice of ham + 2 slices of cheese + 1 piece of lettuce = 1 sandwich

Figure 1. How many sandwiches you are able to make depends on which ingredient you run out of first.

BACKGROUND

Referring to the sandwich example from the INTRODUCTION, we have certain amounts of the ingredients: three loaves of bread, four packages of cheese, six packages of ham, and three heads of lettuce. Given these specific quantities, we must determine which ingredient limits the number of sandwiches that can be produced. Because we know the specific ratio of ingredients per sandwich, the total number of sandwiches made from each ingredient can be calculated as a series of dimentional analysis steps:

$$\text{3 loaves of bread X } \frac{\text{24 slices}}{\text{1 loaf}} \text{ X } \frac{\text{1 sandwich}}{\text{2 slices of bread}} = \text{36 sandwiches}$$

$$\text{4 packages of cheese X } \frac{\text{16 slices of cheese}}{\text{1 package}} \text{ X } \frac{\text{1 sandwich}}{\text{2 slices of cheese}} = \text{32 sandwiches}$$

$$\text{6 packages of ham X } \frac{\text{8 slices of ham}}{\text{1 package}} \text{ X } \frac{\text{1 sandwich}}{\text{1 slice of ham}} = \text{48 sandwiches}$$

$$\text{3 heads of lettuce X } \frac{\text{14 pieces of lettuce}}{\text{1 head of lettuce}} \text{ X } \frac{\text{1 sandwich}}{\text{1 piece of lettuce}} = \text{42 sandwiches}$$

From our calculations we determine that the cheese is the ingredient (reactant) that yields the fewest sandwiches (lower amount of product). In chemical terminology, the cheese is the "limiting reactant". After the thirty-two possible sandwiches are made, the cheese is used up and there are excess quantities of bread, ham and lettuce.

Limiting Reactants

The example of producing sandwiches provides a non-chemical analogy for the chemical concept of limiting reactants. Until this point in your study of chemistry, you were typically given one reactant and asked to calculate the amount of product produced assuming that the other reactants were in excess. In those types of problems, you were actually being given the limiting reactant, without actually calling it by that name. Now you will learn what to do if you are given the masses of each reactant and you are required to determine which is the limiting reactant. In general, limiting reactant problems can be solved using the flowchart in Figure 2. The number of steps required for our work depends on our actual starting and ending places in the flow chart.

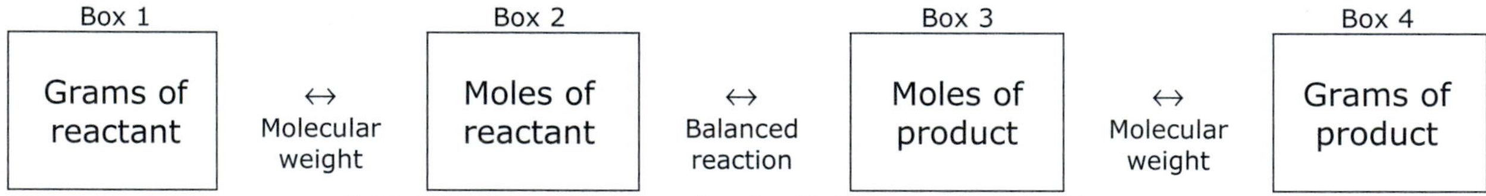

Figure 2. Flowchart for performing limiting reactant problems.

Sample problem: You are given 2.0 grams of zinc and 2.5 grams of silver nitrate and told that they react with each other as in Equation 1. What is the limiting reactant? How many grams of silver can be formed?

$$Zn\ (s) + 2\ AgNO_3\ (aq) \rightarrow 2\ Ag\ (s) + Zn(NO_3)_2\ (aq) \qquad \text{Equation 1}$$

Solution to Sample problem: As in the example of the sandwiches, in order to solve this problem, we must identify which species (Zn or $AgNO_3$) limits how much chlorine gas we can make. Table 1 shows a method for solving limiting reactant problems. The method is broken down into 4 steps. The column on the left describes what is involved in each step and the column on the right shows the actual work for the sample problem. With a little practice these steps can be generalized and used to solve most dimensional analysis problems.

Table 1: Solution to Sample Problem

Step 1: Organize the problem	
Identify what is **known**.	**Known:** $Zn(s) + 2AgNO_3(aq) \rightarrow 2Ag(s) + Zn(NO_3)_2(aq)$ Mass(Zn)=2.0 grams Mass($AgNO_3$)=2.5 grams
Identify **unknown**.	**Unknown:** Limiting reactant? Mass of Ag?
Make **prediction**.	**Prediction:** We have fewer grams of Zn, but the molar mass of Zn is so much less, that it probably corresponds to more moles than $AgNO_3$. That coupled with the fact that each mole of Zn generates two moles of Ag, suggests that the Zn is in excess and the $AgNO_3$ is the limiting reactant.
Step 2: Outline your problem using the flowchart in Figure 2	
Determine starting and ending points on the flow chart. Use these to plan an **outline** for your calculation.	**Outline:** We are given the mass of Zn provided (corresponds to "Box 1" on the flow chart). This will be converted to moles of Zn ("Box 2") using the molar mass. The moles of N_2 are then converted to moles of Ag ("Box 3") using the stoichiometry from the balanced reaction. The number of grams of Ag ("Box 4") is then determined. The entire process is then repeated using the mass of $AgNO_3$ in place of the Zn. The species that produces the smallest mass of Ag is the limiting reactant.
Identify the appropriate **connections** to carry out your outline.	**Connections:** **Assuming Zn is limiting reactant.** • To convert from mass of Zn to moles of Zn, ("Box 1" to "Box 2") use: $\frac{65.39\text{ g Zn}}{\text{mole Zn}}$ • To convert from moles of Zn to moles of Ag, ("Box 2" to "Box 3") use: $\frac{1\text{ mole of Zn}}{2\text{ mole of Ag}}$ • To convert from moles of Ag to grams of Ag, ("Box 3" to "Box 4") use: $\frac{107.9\text{ g Ag}}{\text{mole Ag}}$ **Assuming $AgNO_3$ is limiting reactant.** • To convert from mass of $AgNO_3$ to moles of $AgNO_3$, ("Box 1" to "Box 2") use: $\frac{169.9\text{ g }AgNO_3}{\text{mole }AgNO_3}$ • To convert from moles of $AgNO_3$ to moles of Ag, ("Box 2" to "Box 3") use: $\frac{2\text{ moles of }AgNO_3}{2\text{ mole of Ag}}$ • To convert from moles of Ag to grams of Ag, ("Box 3" to "Box

	4") use: $\frac{107.9 \text{ g Ag}}{\text{mole Ag}}$
Step 3: Perform steps in outline making sure the units cancel.	
At this point you can either do all the steps as one, continuous dimensional analysis problem or perform each of the steps individually as shown in the example.	**Assuming Zn is limiting reactant.** • Convert from mass of Zn to moles of Zn using the connection found in Step 2: $2.0 \text{ g Zn} \bullet \frac{1 \text{ mol Zn}}{65.39 \text{ g Zn}} = 0.0306 \text{ moles Zn}$ • Convert from moles of Zn to moles of Ag: $0.0306 \text{ moles Zn} \bullet \frac{2 \text{ mole of Ag}}{1 \text{ mole of Zn}} = 0.0612 \text{ moles Ag}$ • Convert from moles of Ag to grams of Ag: $0.0612 \text{ moles of Ag} \bullet \frac{107.9 \text{ g Ag}}{\text{mole Ag}} = \mathbf{6.6 \text{ g Ag}}$ **Assuming $AgNO_3$ is limiting reactant.** • Convert from mass of $AgNO_3$ to moles of $AgNO_3$ using the connection found in Step 2: $2.5 \text{ g } AgNO_3 \bullet \frac{1 \text{ mol } AgNO_3}{169.9 \text{ g } AgNO_3} = 0.0147 \text{ moles } AgNO_3$ • Convert from moles of $AgNO_3$ to moles of Ag: $0.0147 \text{ moles } AgNO_3 \bullet \frac{2 \text{ moles of Ag}}{2 \text{ mole of } AgNO_3} = 0.0147 \text{ moles Ag}$ • Convert from moles of Ag to grams of Ag: $0.0147 \text{ moles of } AgNO_3 \bullet \frac{107.9 \text{ g Ag}}{\text{mole Ag}} = \mathbf{1.6 \text{ g Ag}}$ **Because the $AgNO_3$ makes less Ag, the $AgNO_3$ is the limiting reactant. A total of 1.6 g of Ag are made.**
Step 4: Check your answer	
Correct **significant figures** and **units**?	Yes, the answer has two significant figures.
Is the **question answered**?	Yes, the question asked to find the limiting reactant and the mass of Ag.
Does the result **make sense**?	The answer agrees with the initial prediction. We expected the $AgNO_3$ to be the limiting reactant.

The amount of Ag calculated in the Sample problem is an example of a theoretical yield. The theoretical yield is the maximum product that can be expected if 100% of the limiting reactant is converted successfully to products. The theoretical yield is contrasted with the actual yield which is the real amount of the product that is made experimentally. If we carry out the reaction described in the Sample problem and we obtain an actual yield of 1.4 g of Ag, the percent yield is calculated using Equation 2.

$$\% \text{ yield} = \frac{\text{actual yield}}{\text{theoretical yield}} \times 100 = \frac{1.4 \text{ g Ag}}{1.6 \text{ g Ag}} \times 100 = 88\ \% \qquad \text{Equation 2}$$

OVERVIEW

In this experiment, you will combine sulfuric acid and aqueous barium chloride to produce a precipitate, barium sulfate and hydrochloric acid. The precipitate will be isolated by filtration and the theoretical yield will be calculated. You will predict the limiting reactant, and then verify your hypothesis in the lab.

PROCEDURE

Chemicals Used	Materials Used
0.20 M Barium chloride 0.60 M Sulfuric acid 1 M Hydrochloric acid Acetone	Heat resistant gloves Various pipets and pipet bulb 50-mL Ehrlenmeyer flasks (3) Glass stirring rod Hot plate Analytical balance Buchner funnel and filter paper Vacuum filtration apparatus (Figure 3) Ring stand and clamp Oven

1. Working in pairs, students will be assigned to work with one of the following volumes of 0.20 M barium chloride: 5, 10, 15, 20, 25 or 30 mLs. Two pairs of students will be assigned the same volume for comparison.

2. Pipet your assigned volume of 0.20 M barium chloride into a 50-mL flask. Warm the flask in the hood on a hot plate to near boiling.

3. Rinse your pipet with distilled water. Pipet 5.00 mL of 0.60 M sulfuric acid into another 50-mL flask. Again, rinse your pipet and add 5.00 mL of 1 M hydrochloric acid to the same flask. Warm this flask on a hot plate to near boiling.

4. While the solutions are heating, weigh a piece of filter paper and place it in the funnel. Moisten the filter paper with de-ionized water so that it adheres to the funnel. Place the funnel with the filter paper into the filtration apparatus with a rubber tube attached to a vacuum (Figure 3).

Figure 3. Vacuum filtration apparatus.

5. Work in the hood with heat resistant gloves. When both solutions are hot, slowly (while stirring) add the barium chloride to the flask with the sulfuric acid. Rinse any precipitate off the stirring rod into the flask with de-ionized water. Heat the solution for another 15 minutes to ensure formation of large crystals of precipitate. Add an additional 5 mLs of water as needed to prevent total evaporation. Allow the mixture to cool slowly to room temperature.

6. Pour the mixture onto the filter paper in the funnel. Rinse the residual precipitate into the funnel using de-ionized water. Turn on the vacuum for 3 minutes. Once the precipitate is fairly dry, add about 5 mLs of acetone to the funnel to speed the drying. When the filter paper is dry, remove the paper and precipitate, and determine the mass of the paper and precipitate. Alternatively, the filter paper may be dried in a laboratory oven.

7. Before leaving, calculate the actual mass of barium sulfate that you isolated and add your results (averaged with the other group that use the same volume of barium chloride) to the blackboard for the class graph of "mass of barium sulfate made versus volume of barium chloride used".

Limiting Reactants: How Much $BaSO_4$ Can We Make?

Name:	Lab Instructor:
Date:	Lab Section:

PRE-LABORATORY EXERCISES

1. Write and balance the equation for the reaction you will be running as described in the OVERVIEW.

2. According to the concentrations in the PROCEDURE and assuming that you are assigned to work with 1 mL of barium chloride, calculate the theoretical number of grams of barium sulfate that should be produced. What is the limiting reactant? Write up your answer following the steps in Table 1 in the BACKGROUND.

OVER →

PRE-LABORATORY EXERCISES continued...

3. You allow 75.0 grams of carbon monoxide to react with 58.0 grams of hydrogen according to the following process. Determine the limiting reactant and the amount of methanol produced. Again, show your work following the steps in Table 1.

$$CO\ (g) + H_2\ (g) \rightarrow CH_3OH\ (l)$$

Limiting Reactants: How Much $BaSO_4$ Can We Make?

Name:	Lab Instructor:
Date:	Lab Section:

RESULTS and POST-LABORATORY QUESTIONS

Assigned volume of barium chloride	
Actual mass of barium sulfate isolated	

1. Calculate the theoretical yield of barium sulfate using your assigned volume.

2. Calculate the percent yield of barium sulfate.

3. Attach a copy of the graph of the average class data for "mass of barium sulfate made versus volume of barium chloride used". What is occurring before the graph levels off? What is indicated after the graph levels off?

Electrolytes in Solution: Completing the Circuit

Kristen Spotz

OBJECTIVES

- Understand how various compounds behave when dissolved in water.
- Develop a method for categorizing compounds based on the ability of their aqueous solutions to conduct electricity.
- Practice naming some common compounds.

INTRODUCTION

Imagine taking a walk along the seashore on a clear sunny day. As you enter the water, you think about the chemistry you have learned. You know that the oceans of our world are far from being pure H_2O. They are, in fact, aqueous solutions composed of countless substances dissolved in the water. Do these dissolved substances exist in solution as neutral molecules or as charged ions having the capacity to conduct an electrical current? What if the sky suddenly darkened and a bolt of lightening struck directly above you? (Figure 1) Are you safe, or have you just become part of a gigantic electrical circuit?

Figure 1. A bolt of lightening crashes off shore

BACKGROUND

Water and Solubility

Water is a remarkable solvent. Aqueous solutions can be made by dissolving in water many different compounds, both ionic and covalent/molecular. The oceans and lakes around us, as well as our very cells, are all examples of aqueous solutions.

As you might expect, not all things dissolve in water to the same extent. Scientists quantify the extent to which a particular solute dissolves in water by determining the solubility of the solute at a given temperature (typically, grams of solute per liter of water). For example, NaCl is very soluble in water. At 20°C, one liter of water can accommodate 365 g of NaCl before becoming saturated. Other compounds, such as $CaSO_4$, are labeled as insoluble meaning that less than 1 g of the material will dissolve in one liter of water. Insoluble compounds require very little solute before becoming saturated.

It is important to note that although textbooks give solubility rules describing compounds as either strictly soluble or insoluble, in actuality, solubility covers a whole spectrum. Even soluble compounds have their limits. Once the solubility of a compound is reached, no more will dissolve. For example, if you attempt to add more than 365 grams of NaCl to 1 liter of water at 20°C, the excess will sit on the bottom in the form of an undissolved solid. On the opposite end of the spectrum, even "insoluble" compounds can dissolve to a very small extent.

Physical Properties of Aqueous Solutions

Depending on the nature of the dissolved solute (ionic or covalent/molecular), the solute may be present in solution in the form of ions or as neutral molecules. All soluble ionic compounds (this includes the strong bases) exist as ions in solution. For example, when dissolved in water, NaCl exists as Na^+ and Cl^- ions (Figure 2). Even in the case of slightly soluble ionic compounds, where the majority of the solute is left undissolved on the bottom, all of the compound that did dissolve is in the form of ions. Although Figure 2 serves as an adequate representation of the behavior of soluble ionic compounds in water, a more realistic drawing is presented in Figure 3, where the dissolved ions are shown to be hydrated.

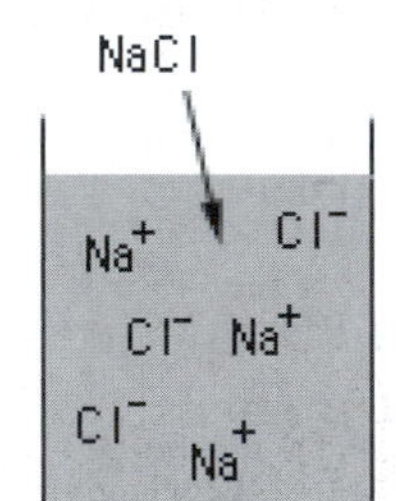

Figure 2. The behavior of ionic compounds (aq).

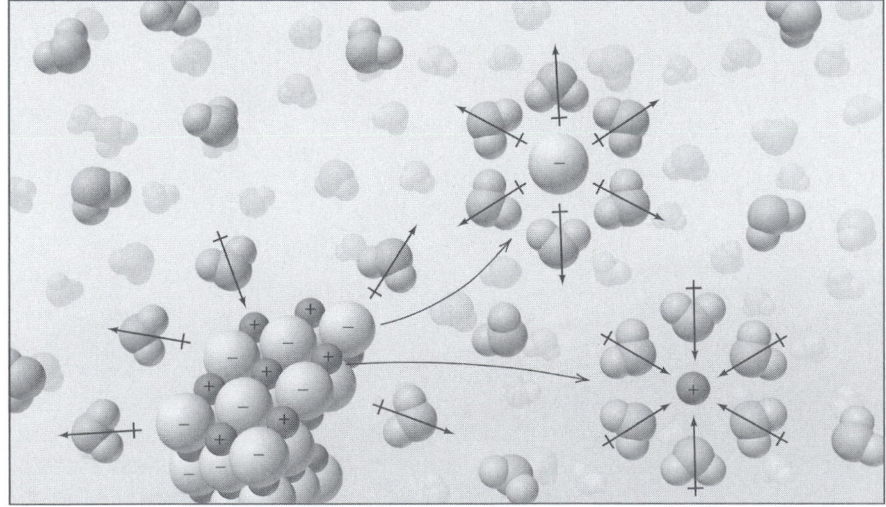
Figure 3. Hydrated Na^+ and Cl^- ions.

If, however, the solute placed in water is covalent/molecular, then the behavior of the solute is more complicated. Sometimes the solution will consist of neutral molecules as is the case with O_2 (Figure 4). Other covalent compounds break apart in solution to create varying amounts of ions. Let's take a closer look at this last category of compounds.

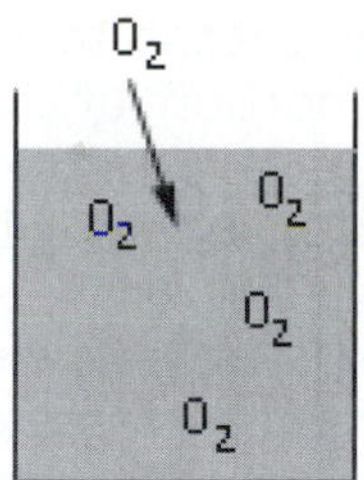

Figure 4. The behavior of nonelectrolytes (aq)

Strong and weak acids along with weak bases are a small, but important group of molecules that ionize in water. The degree of ionization depends on the strength of the acid or base. For example, strong acids, such as HCl, ionize nearly 100% in water to produce H^+ and Cl^- ions (Figure 5). However, with weak acids and weak bases, such as HF or NH_3 only a small fraction (less than 5%) of the molecules ionize leaving the majority of the solute molecules intact (Figure 6).

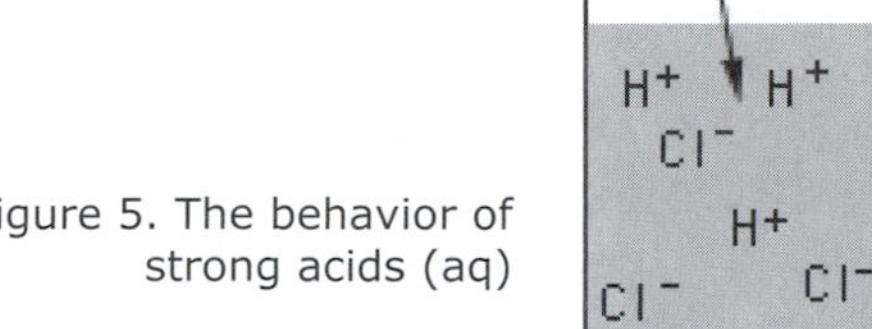

Figure 5. The behavior of strong acids (aq)

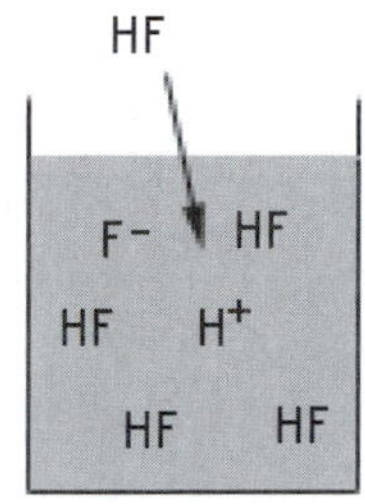

Figure 6. The behavior of weak acids and bases (aq)

Though it is not crucial to this discussion, you may be aware that free H^+ ions don't actually exist in solution. It is more realistic to represent them as bonded to a water molecule in the form of hydronium ions.

Completing the Circuit

In 1883, Svante Arrhenius proposed that ions are responsible for the conductivity of electricity through a solution. So, one way to identify the form (ionic or covalent) of the dissolved solute is to test the electrical conductivity of the particular aqueous solution (Figure 7a). Because the flow of electricity requires the presence of mobile charged particles, aqueous solutions that contain ions will conduct electricity (Figure 7b) and are therefore called electrolytes.

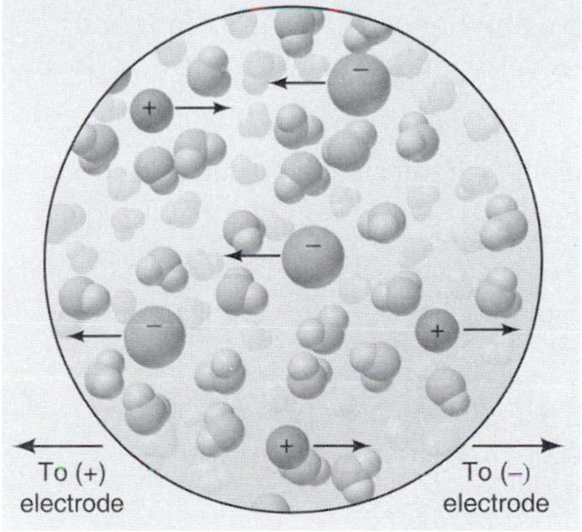

Figure 7b. In solution, anions move towards the positive electrode and cations move towards the negative electrode.

Figure 7a. Conductivity apparatus

The extent to which a particular solution conducts an electrical current depends on what proportion of solute particles dissociate into ions (Figure 8). Compounds whose solutions contain a large fraction of the solute in the form of ions (all soluble ionic compounds or strong acids) will conduct a large current and are called strong electrolytes. By contrast, a weak electrolyte is a solution that contains only a relatively small fraction of ionized solute particles (the weak acids and weak bases). As a result, solutions of weak electrolytes only conduct a small amount of current. Finally, a solution of a non-electrolyte, for example CO_2, contains only dissolved molecules and will not conduct an electrical current.

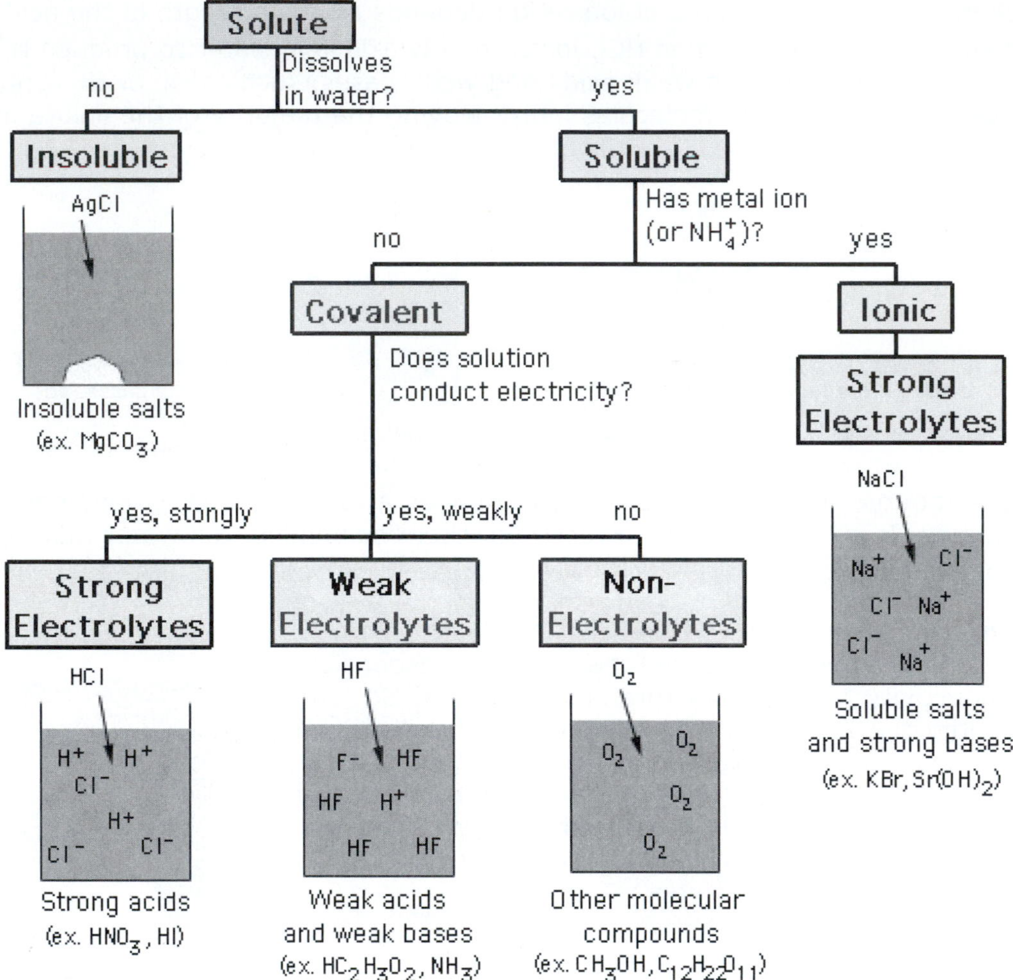

Figure 8. Flowchart for determining whether a compound is a strong, weak, or non-electrolyte.

OVERVIEW

In this experiment, you will determine the relative solubility of several ionic compounds by adding the compound to water in small portions until no more dissolves. You will also use a conductivity apparatus (Figure 7a) to test the conductivity of several solutions.

PROCEDURE

Part A. Relative Solubility

Chemicals Used	Materials Used
KNO_3 (s) $CaCO_3$ (s) $NaC_2H_3O_2$ (s)	50 mL Beakers (3) Glass stirring rod 50 mL Graduated cylinder Hot plate Thermometer

Label three 50-mL beakers "KNO_3", "$CaCO_3$" and "$NaC_2H_3O_2$". Using a graduated cylinder, add 20.0 mL of de-ionized water to each of the beakers. Add 1.0 gram of each solid to the appropriately labeled beaker. Stir. Continue adding additional 1.0-gram portions to the corresponding beakers until the solutions are saturated or until you have added 5 grams, which ever comes first. Record the number of grams of each solid that you were able to dissolve.

Heat 20.0 mL of de-ionized water to around 60°C. Determine the solubility of KNO_3 in water at 60°C. Discard waste solutions according to your instructor's directions.

Part B. Strong, Weak and Non-Electrolytes

Chemicals Used	Materials Used
NaCl (s) NaCl (1M, aq) $C_{12}H_{22}O_{11}$ (1M, aq) HCl (1M, aq) $HC_2H_3O_2$ (1M, aq) NaOH (1M, aq) NH_3 (1M, aq)	Conductivity apparatus

CAUTION: Be careful to avoid an electric shock when working with the conductivity apparatus. Unplug the conductivity apparatus before rinsing the electrodes or putting the electrodes down.

Your instructor will have stations with labeled 100-mL beakers containing 50 mL of each of the chemicals listed above. Test the conductivity of each sample and record the relative brightness of the light using the following scale: bright, dim, faint or no light. Be sure to rinse the electrodes before testing each solution.

Part C. Effect of Concentration on Conductivity

Chemicals Used	Materials Used
2 M HCl	100-mL beaker 50-mL Graduated cylinder Plastic pipet Conductivity apparatus

Using the graduated cylinder, transfer 50.0 mL of de-ionized water to a 100-mL beaker. Add 2 M hydrochloric acid drop-wise (up to a maximum of 50 drops) to the de-ionized water. Stir the solution and observe the conductivity after each drop of hydrochloric acid is added. Record the number of drops at which you first notice the light and again when the light is bright. Discard waste solutions according to your instructor's directions.

Part D. Conductivity of Household Substances

Chemicals Used	Materials Used
Epsom salt ($MgSO_4$) Rubbing alcohol (C_3H_7OH) Baking soda ($NaHCO_3$) Vinegar ($HC_2H_3O_2$)	100-mL Beakers (4) 50-mL Graduated cylinder Stirring rod Conductivity apparatus

Your instructor will have stations with labeled 100-mL beakers containing 50 mL of each of the household chemicals listed above. Test the conductivity of each sample and record the relative brightness of the light using the following scale: bright, dim, faint or no light. Be sure to rinse the electrodes before testing each solution.

Electrolytes in Solution: Completing the Circuit

Name:	Lab Instructor:
Date:	Lab Section:

PRE-LABORATORY EXCERCISES

1. Define the underlined words in the BACKGROUND section.

2. In this laboratory experiment, you will be studying the solubility of KNO_3, $CaCO_3$, and $NaC_2H_3O_2$ in water. Use the solubility rules in your textbook to make predictions about whether each compound is expected to be soluble or insoluble in water.

3. Which sample of water (seawater, tap water or de-ionized water) do you think will be the best conductor of electricity? The poorest conductor? Explain.

4. Why is it important to use de-ionized water when preparing the aqueous solutions in this experiment?

OVER →

PRE-LABORATORY EXCERCISES continued...

5. For each compound (calcium chloride, glucose and sodium iodide) give the formula, state whether it is ionic or covalent, whether it is soluble in water, whether it forms ions in water, and whether an aqueous solution conducts electricity. Also make a drawing illustrating the behavior of the compound in water. Use Figure 8 to help you make your predictions. Use the space below to arrange your answers in the form of a table.

Electrolytes in Solution: Completing the Circuit

Name:	Lab Instructor:
Date:	Lab Section:

RESULTS and POST-LABORATORY QUESTIONS

Part A. Relative Solubility

Compound	Name	Grams Dissolved (room temp. water)	Grams Dissolved (warm water)
KNO_3			
$CaCO_3$			
$NaC_2H_3O_2$			

Use your data from Part A to calculate the solubility of each of the compounds in room temperature water. Report the solubility in both grams/mL and molarity.

Which of the compounds would you describe as very soluble? Slightly soluble? Insoluble? How do your results compare to those predicted from the solubility rules listed in your text (see PRE-LABORATORY EXERCISE #2)?

What effect did the warm water have on the solubility of the KNO_3? Can you think of a practical example of this behavior?

Part B. Strong, Weak and Non-Electrolytes

Compound	Name	Relative Brightness	Strong, Weak or Non-Electrolyte
NaCl (s)			
NaCl (aq)			
$C_{12}H_{22}O_{11}$ (aq)	sucrose		
HCl (aq)			
$HC_2H_3O_2$ (aq)			
NaOH (aq)			
NH_3 (aq)			

Referring to Part B, why did the bulb light up when placed in an aqueous solution of sodium chloride, but not in solid sodium chloride?

OVER →

RESULTS and POST-LABORATORY QUESTIONS continued...

Part C. Effect of Concentration on Conductivity

From Part C, how many drops of HCl were added before you first noticed a light? How many drops before the light became bright?

Based on your results, if you placed one gram of a soluble ionic compound (ex: NaCl) in a full-size swimming pool filled with de-ionized water, would you expect the solution to conduct an electrical current? Explain.

Part D. Conductivity of Household Substances

Compound	Name	Relative Brightness	Strong, Weak or Non-Electrolyte
$MgSO_4$ (aq)			
C_3H_7OH (aq)	propanol		
$NaHCO_3$ (aq)			
$HC_2H_3O_2$ (aq)			

In Parts B and D, you determined whether NaOH, NH_3, C_3H_7OH and $NaHCO_3$ are strong, weak or non-electrolytes. Based on your results, draw beakers showing atomic scale representations of aqueous solutions of these compounds. Use the drawings in Figure 8 as examples.

If You Are Not Part of the Solution: Precipitation Reactions

Jeffrey Paradis

OBJECTIVE

- Use solubility rules to predict the products of precipitation reactions.
- Draw atomic scale pictorial representations of precipitation reactions.
- Explore a practical application of precipitation reactions: determine the ions present in an unknown solution using qualitative analysis.

INTRODUCTION

The Aquaphor Water Co. is currently experiencing some problems with their water purification plant. They have detected some metal cations in the drinking water, but have not determined the identity of the metals. They have, however, narrowed it down to the following metal cations: Ag^+, Zn^{2+},Ca^{2+}. Aquaphor Water Co. has sent you a sample of their contaminated water and will pay you handsomely if you can identify which metal cation(s) are present.

BACKGROUND

Precipitation Reactions

We know from experience that not all solutes are soluble in all solvents. Whether or not an ionic compound will dissolve in water is summarized in the Solubility rules found in most general chemistry textbooks. Interestingly, these rules hold true even if the ions come together from different soluble sources. This behavior is descriptive of an important class of chemical reactions called precipitation reactions.

As an example, consider the reaction between aqueous solutions of NaI and $Pb(NO_3)_2$. When the $Pb(NO_3)_2$ (aq) is added to the NaI (aq) a yellow solid, the precipitate, is immediately formed (Figure 1).

Figure 1. The reaction of $Pb(NO_3)_2$ (aq) with NaI (aq)

To figure out the identity of the solid, it is helpful to break the process down into steps and to draw a picture representing what is occurring on the atomic scale. Begin by writing the two reactants (Equation 1).

$$NaI\ (aq) + Pb(NO_3)_2\ (aq) \rightarrow \qquad \text{Equation 1}$$

Next we switch each anion and pair it with the other cation. This is often called a double-displacement or metathesis reaction. In this case, we end up with $NaNO_3$ and PbI_2. We then consult the solubility rules in our textbook to see if either, neither or both of these compounds is soluble in water. All nitrates are soluble. Therefore the $NaNO_3$, as well as the original $Pb(NO_3)_2$, are soluble in water. Most iodides are soluble, like the NaI, but there are a few important exceptions. When the Pb^{2+} ions come into contact with the I^- ions, the precipitate PbI_2 is formed. The colorless Na^+ and NO_3^- ions remain floating in the solution. The states of the products can now be indicated (Equation 2).

$$NaI\ (aq) + Pb(NO_3)_2\ (aq) \rightarrow NaNO_3\ (aq) + PbI_2\ (s) \qquad \text{Equation 2}$$

Balancing the reaction results in Equation 3.

$$2NaI\ (aq) + Pb(NO_3)_2\ (aq) \rightarrow 2NaNO_3\ (aq) + PbI_2\ (s) \qquad \text{Equation 3}$$

It is often useful to write the compounds as they appear in solution. In other words, if an ionic compound is aqueous and dissolved in solution, we write it broken up into the individual ions. If an ionic compound is in the form of a solid, we leave it alone. This form of a chemical equation is called the total ionic equation (Equation 4).

$$2Na^{+}\,(aq) + 2I^{-}\,(aq) + Pb^{2+}\,(aq) + 2NO_3^{-}\,(aq) \rightarrow 2Na^{+}\,(aq) + 2NO_3^{-}\,(aq) + PbI_2\,(s)$$

Equation 4

Note that the above total ionic equation is still balanced and that only the PbI_2 precipitate is indicated as not existing as ions floating in solution.

Lastly, many students find it conceptually helpful if they are able to visualize the process by drawing a picture representing the reaction on the atomic scale (Figure 2). The two boxes on the left represent the nature of the two reactants before they are mixed. The box to the right of the arrow indicates that the PbI_2 has settled to the bottom as a solid while the Na^+ and NO_3^- ions remain in solution.

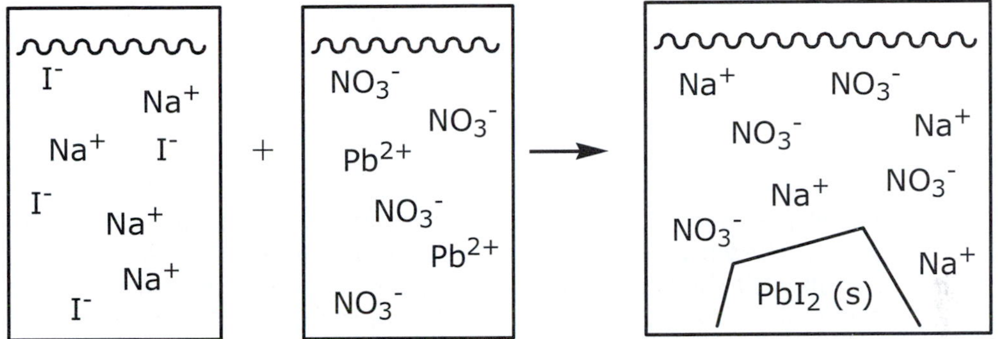

Figure 2. Atomic scale pictorial representation of the precipitation reaction between aqueous solutions of NaI and $Pb(NO_3)_2$. Water molecules have not been included for clarity.

Qualitative Analysis

Qualitative analysis is a technique used to separate and identify ions in a mixture. This type of research utilizes a variety of skills and knowledge. It is imperative that you are familiar with the solubility rules. The key to discovering the ion(s) present is to pick a set of conditions that causes a single ion or small subset of ions to precipitate while leaving all the other ions in solution. Figure 3 shows a general qualitative analysis scheme for separating ions that are in solution. After each time the sample is centrifuged, the liquid is poured off or removed with a pipet to separate it from the precipitate.

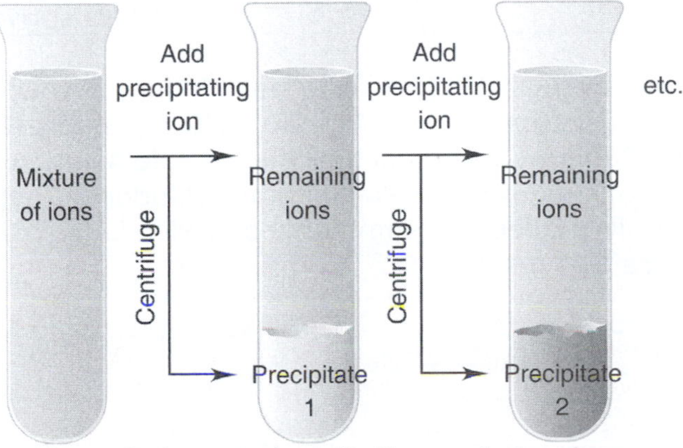

Figure 3. A general qualitative analysis scheme

OVERVIEW

In this experiment, you will use precipitation reactions to identify an unknown solution (Part A) and you will go to work for the Aquaphor Water Co. using qualitative analysis to help them solve a problem with their water purification plant (Part B).

PROCEDURE

Part A. Identifying an Unknown

Chemicals Available		Materials Available
Sample Solutions	Precipitating Solutions	24-wellplate (2)
$AlCl_3$	H_2SO_4	
$AgNO_3$	K_2CrO_4	
Na_2CO_3	$Cu(C_2H_3O_2)_2$	
NH_4OH	KOH	
HCl	$Pb(NO_3)_2$	
Unknown solution	Na_2HPO_4	

Set up two 24 well-plates next to each other forming a matrix of eight wells across and six wells down. Begin with a Sample solution and fill each of the six wells in the first column 1/3 full with the Sample solution. Then use the Precipitating solutions to fill each of these six wells until 2/3 full with a different Precipitating solution. Carefully record which solution is used in each row and each column as well as your observations of what happened in each well. Continue this procedure, using the Precipitating solutions on the remaining Sample and the Unknown solutions.

Part B. Qualitative Analysis

Chemicals Available		Materials Available
Sample solutions	Precipitating solutions	Centrifuge
0.10 M $Zn(NO_3)_2$	0.10 M NaOH	Droppers
0.10 M $Ca(NO_3)_2$	0.10 M NaCl	24-well plate
0.10 M $AgNO_3$	0.10 M Na_2CO_3	Centrifuge tubes (2)
Unknown Solution	0.10 M $NaC_2H_3O_2$	Test tube rack
		Glass stirring rod

Go back and read the INTRODUCTION for a description of your task for Part B and the Qualitative Analysis section of the BACKGROUND for a general procedure on how to separate the ions. Begin by using each of the Sample solutions to verify which will precipitate with each of the provided Precipitating solutions. These results should be compared to the solubility rules you summarized in PRE-LABORATORY EXERCISE #1. Record your procedure and results.

If You are Not Part of the Solution: Precipitation Reactions

Name:	Lab Instructor:
Date:	Lab Section:

PRE-LABORATORY EXERCISES

1. You will need to use solubility rules when performing the qualitative analysis part of this experiment. Use your textbook to summarize the solubility rules for the following possible precipitation solutions: OH^-(Hint: the strong bases are soluble), Cl^-, CO_3^{2-}, and $C_2H_3O_2^-$.

2. Following the model from the BACKGROUND section, answer the questions below for each of the two reactions:

   Reaction #1: $Cu(C_2H_3O_2)_2$ (aq) + Na_2CO_3(aq) →

   a. Predict the products:

   b. Balance the reaction:

   c. Write the total ionic equation:

   d. Draw a picture representing the reaction on the atomic scale:

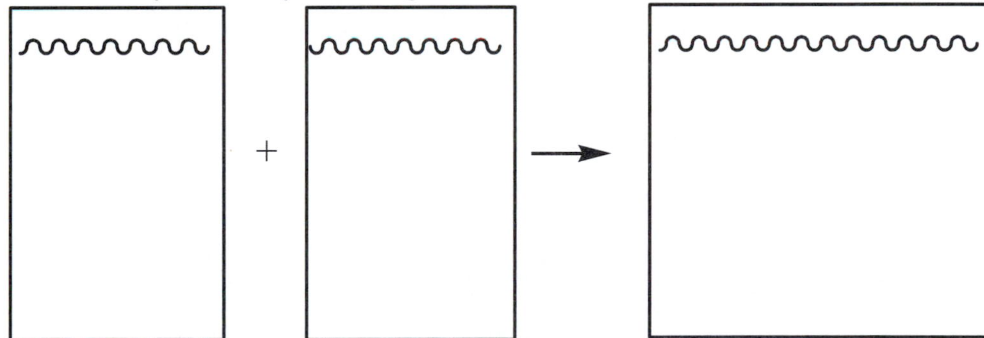

OVER →

PRE-LABORATORY EXERCISES continued...

Reaction #2: KOH (aq) + $AlCl_3$ (aq) →

a. Predict the products:

b. Balance the reaction:

c. Write the total ionic equation:

d. Draw a picture representing the reaction on the atomic scale:

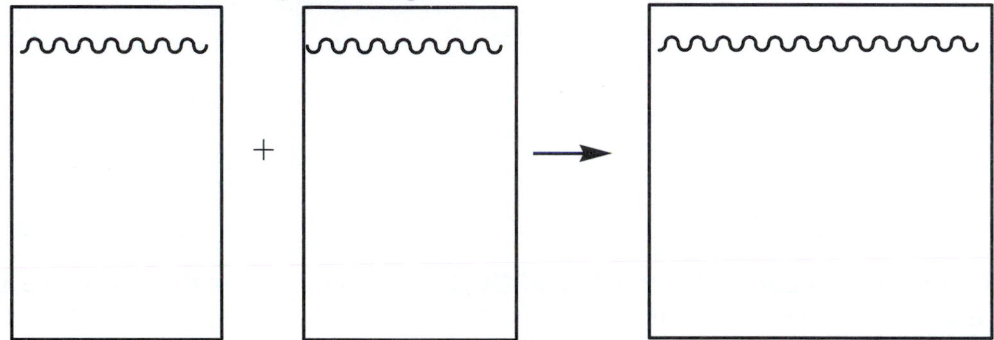

If You are Not Part of the Solution: Precipitation Reactions

Name:	Lab Instructor:
Date:	Lab Section:

RESULTS and POST-LABORATORY QUESTIONS

Part A. Identifying an Unknown

Use the blank matrix below to fill in your results. The shaded areas should be filled with the formula of the Sample, Precipitating or Unknown solution used in that column or row. The empty boxes should indicate your observations for that test.

What was the code for your unknown in Part A?

What was the identity of your unknown? Briefly explain how you came to that conclusion.

Two of the reactions you carried out in Part A were reactions you discussed in PRE-LABORATORY EXERCISE #2. Did the results from Part A correspond, in general terms, with your predictions? Explain.

Given the chemical formulas, provide the corresponding name for each Sample solution and Precipitating solution you worked with in Part A.

OVER →

RESULTS and POST-LABORATORY QUESTIONS continued...

Part B. Qualitative Analysis

Write up the procedure you followed for Part B. Include a comparison of the results of the test of your Sample solutions with the answers you found for PRE-LABORATORY EXERCISE #1. Also, be sure to indicate the code for your unknown, what cation(s) were present in your unknown and why you think they were present.

An Introduction to Oxidation-Reduction Chemistry: The Breath Analyzer

Joel Kelner

OBJECTIVE

- Master the terminology used in oxidation-reduction reactions.
- Use oxidation numbers to monitor the transfer of electrons during a chemical reaction.
- Make a working model of an alcohol breath analyzer.

INTRODUCTION

Drunk driving is a national problem. Across the country, over 17,000 people are killed annually in car accidents involving alcohol. In fact, someone dies every 30 minutes in an alcohol-related accident.

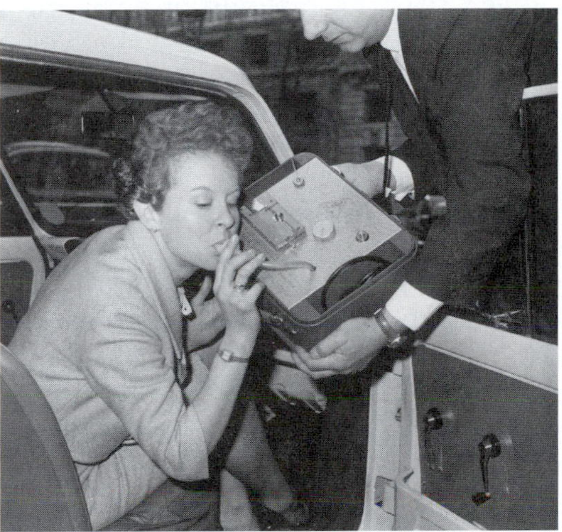

Figure 1. Chemical tests for the presence of alcohol on a person's breath are a practical example of an oxidation-reduction reaction.

In an effort to help remove intoxicated individuals from the road, scientists have developed a simple and reliable test to monitor an individual's blood alcohol content (BAC) (Figure 1). If you take a look at your driver's license, you will most likely see a phrase to the effect that "operation of a motor vehicle constitutes consent to any sobriety test required by law." In most states, refusing to take the tests, whether eventually convicted in court or not, carries the same punishment as a guilty ruling.

What is a chemical sobriety test and how does it work?

BACKGROUND

Sobriety Tests

There are many different types of tests that can be administered to an individual suspected of intoxication. Non-chemical tests include verbal tests, coordination tests, and observation of dilated pupils and can give reasonable suspicion to further administer chemical tests. While non-chemical tests are possible qualitative indicators for the presence of alcohol, chemical tests present the hard proof that is needed to bring a drunk driver to trial. Most chemical tests provide quantitative values of the amount of alcohol in the blood. These chemical tests include the common breath analyzers as well as analysis of saliva, perspiration, urine and blood. The different breath analyzers are the most widely used tests for alcohol and while the real "Breathalyzer" test is rather complex, it is based on the same type of oxidation-reduction reaction that we will examine.

How does the breath test work? It is based on the principle that the alcohol in the blood is in equilibrium with alcohol in the lungs. The validity of the breath test relies on the assumption that the alcohol absorbed into the blood is continuously transferred into the lungs and that the makeup of air is the same throughout the lungs and in the exhaled air. The standard Breathalyzer test involves an oxidation-reduction reaction involving potassium dichromate, ethanol and sulfuric acid. To understand the chemistry behind the Breathalyzer test, you must be familiar with basic oxidation-reduction terminology. Begin by using your textbook to answer PRE-LABORATORY EXERCISE #1.

Oxidation Numbers

Now that you've got the basics down, there is a simple way of telling what is oxidized and what is reduced during an oxidation-reduction reaction. Begin by identifying the oxidation number (or oxidation state) of all elements involved. The reactant that exhibits a decrease in its oxidation number (or an increase/gain in the number of electrons) is reduced during the reaction. Likewise, the reactant that displays an increase in oxidation number (or a decrease/loss in the number of electrons) is oxidized. It might help to remember that "**LEO** (**L**ose **E**lectrons **O**xidized) the lion says **GER** (**G**ain **E**lectrons **R**educed)" (Figure 2).

Figure 2. A simple pneumonic for learning oxidation-reduction terminology.

It should be noted that oxidation numbers have no physical significance; rather they allow us to keep track of the movement of electrons. Before you continue, familiarize yourself with the rules for determining oxidation numbers by answering pre-laboratory question #2.

You should have found by now that there are certain elements, like oxygen, that have the same oxidation number in most of their compounds. Other elements, like nitrogen, can take on a variety of oxidation numbers depending on the other elements with which they combine. To figure out the oxidation states in unclear cases, we use the rules to first give us the oxidation numbers we are confident about. We then deduce the oxidation states for the elements that do not follow the rules. For example, consider the nitrate ion, NO_3^-.

(ox. number on N)(# of N's) + (ox. number on O)(# O's) = (charge on species)

4. Let the mixture stand for several minutes while you prepare the clean, 15-cm long piece of plastic tubing. Take a cotton ball and rip it in half. Place half of the cotton ball into the tubing, about two-thirds of the way in. Do not pack it too tightly, as you will eventually want air to pass through the tube.

5. Use a small metal spatula to scoop the acidified potassium dichromate mixture into the non-cotton end of the plastic tubing. Try to insert the mixture into the center of the tube, so you don't get it on the sides. Spread it out so that it covers all around the tube in the middle one-third section.

6. Take the cotton swabs and clean out the area where you put the mixture into the tube. Then, put the other half of the cotton ball into the open side. Again, do not pack the cotton too tightly.

7. Take a balloon and, using a plastic pipette, add about 1 mL (25 drops), of 200 proof ethanol. Fill the balloon with air from an air line in the lab. Be careful not to over-inflate the balloon. Hold the end so the air won't escape and wait about 5 minutes to allow some of the ethanol to vaporize within the balloon. **CAUTION:** Do not blow up the balloon with your mouth. Objects in the laboratory should always be assumed to be contaminated and should never be touched to the face or mouth.

8. Attach the balloon to the end of the tube opposite the one you put the mixture into (the clean end). Release the air and record your observations (Figure 4).

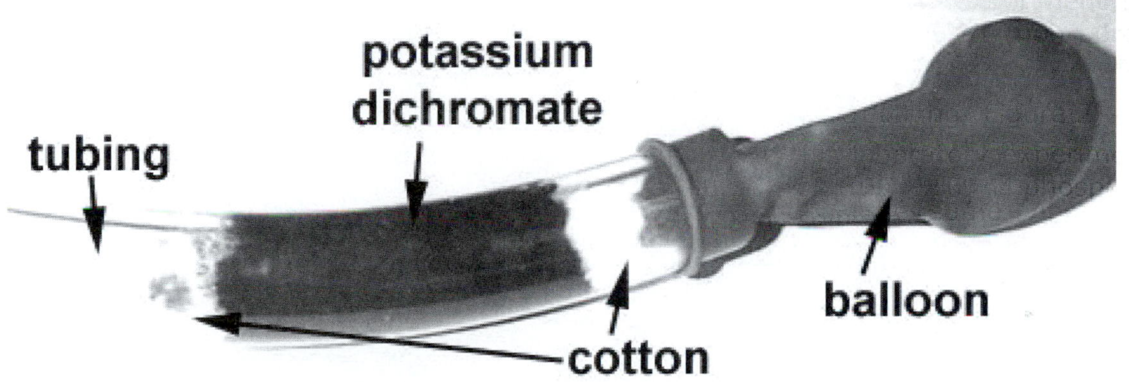

Figure 4. A successfully completed breath analyzer test.

An Introduction to Oxidation-Reduction Chemistry: The Breath Analyzer

Name:	Lab Instructor:
Date:	Lab Section:

PRE-LABORATORY EXERCISES

1. Define the following terms: Oxidation, Reduction, Oxidizing agent, Reducing agent, Oxidation number (or oxidation state).

2. Use the Internet to find the legal blood alcohol content (BAC) level in your state. Record the URL of the web site and the date that you accessed the information.

3. Use your textbook to summarize the rules for determining oxidation numbers.

OVER →

PRE-LABORATORY EXERCISES continued...

4. Use your textbook to find information to help you write chemical reactions for the following three processes: "rusting of iron", "burning magnesium ribbon" and "copper in nitric acid". Label the oxidation number of each species as well as the oxidizing and reducing agents in each of the three reactions.

An Introduction to Oxidation-Reduction Chemistry: The Breath Analyzer

Name:	Lab Instructor:
Date:	Lab Section:

RESULTS and POST-LABORATORY QUESTIONS

1. Using complete sentences, summarize your observations for Part A and B.

2. Write the balance equation for the reaction occurring between the silver nitrate and copper wire in Part A. Label the oxidation numbers for all elements involved. What has been oxidized and what has been reduced?

3. There is some metal floating at the bottom of the beaker in the silver nitrate and copper wire experiment. There is also some metal coating on the wire. What are these metals?

4. Write the balance equation for the reaction occurring in Part B (Equation 3). Label the oxidation numbers for all elements involved in the reaction. What has been oxidized and what has been reduced? What is the oxidizing agent? What is the reducing agent?

Say "Cheese": Cyanotype Photography as an Example of a Photochemical Redox Reaction

Brian Polk

OBJECTIVE

- Gain an understanding of the chemistry involved in the photographic process.
- Use UV radiation to initiate a photochemical redox reaction.
- Capture an image using cyanotype photography.

INTRODUCTION

The cyanotype photographic process was developed by Sir John Herschel in the 1840's. Commercial application followed in the 1870's when architects used this technique to copy drawings in which the image was yellow and the background blue. Thus the term "blueprint" was coined. Applications were seen in the area of traditional photography at the end of the 19th century, but the blue color made it an unfavorable choice for "professional" photographers. Cyanotype photography has seen a resent resurgence in popularity, however, and the picture of "Rackwick at Dusk" (Figure 1) was captured using a cyanotype photographic process. But one question remains, "How is it possible to capture light and to make pictures with chemicals?"

Figure 1: "Rackwick at Dusk" by cyanotype photography. For more images of this type, visit *http://www.mikeware.demon.co.uk/cyanotypes.html*

BACKGROUND

Photography

Photography, as practiced by both amateurs and professionals, is a very diverse science using numerous techniques. Today, most photographic films utilize silver, which, when combined with a heavy halide, proves very useful for photography. In fact, the photographic film you use in your camera is most likely a plastic support with silver bromide microcrystals dispersed in gelatin. In this type of photography, the reaction is catalyzed by visible light. As light interacts with the subject of your picture, different parts of the film containing silver bromide will be exposed to different amounts of light. When more light is present, more silver bromide will react.

In today's laboratory experiment, however, we will look at the lesser known cyanotype photographic process. Cyanotype photography retains the fascinating chemistry of more traditional films, but provides us with a simpler procedure. The cyanotype photographic process is based on the ultraviolet (UV) light initiated redox reaction where iron (III) is reduced to iron (II) which results in a color change of the exposed area of the film.

Redox chemistry

The term "redox" comes from the two words reduction and oxidation. During a redox process an electron is transferred from one species to the other. A typical redox reaction is shown below in Equation 1.

$$Fe\ (s) + Ni(NO_3)_2\ (aq) \rightarrow Fe(NO_3)_2\ (aq) + Ni\ (s) \qquad \text{Equation1}$$

By assigning the oxidation states to each of the species involved in the reaction, one can determine which species is reduced and which is oxidized during the electron transfer. The iron metal has an oxidation state of 0 while the Ni in the nickel(II) nitrate complex is +2. After reaction, the iron has an oxidation state of +2, meaning it lost 2 electrons (was oxidized) while the nickel has an oxidation state of 0, indicating that it had to gain 2 electrons (was reduced).

Photochemistry

As the name implies, photochemistry involves the use of light to perform chemical reactions. Quantum mechanics tells us that electrons have quantized energy levels within atoms. An electron can be excited from one orbital to the next, if the energy provided corresponds to the difference in energy between the two orbitals.

During a photochemical process, light provides energy to the chemicals involved causing the electrons in the chemicals to absorb energy and enter an excited state. From this excited state, the electron will relax to a lower energy state where it is more stable. The electron can relax by either fluorescence or phosphorescence. Another possibility is that the electron could participate in a chemical reaction from the excited state. The higher energy afforded by the excited state typically makes the species more reactive. Therefore, instead of giving off light, the energy is transferred to the chemical reaction.

How does photochemistry relate to photography? By choosing the areas that are exposed to the light, we can choose the areas where the photochemical reactions will occur. Therefore, if we cover certain parts of our photographic "film", the chemicals underneath will remain unreacted and the portions exposed to the UV radiation will take part in the photochemical process.

The exact chemical reactions taking place in cyanotype photography are not fully understood, but the fundamental chemistry is as follows. When iron (*III*) oxalate ion,

$[Fe(C_2O_4)_3]^{3-}$, is exposed to UV light, the iron is reduced to iron(*II*) and some of the oxalate is oxidized to CO_2 (Equation 2). The iron(*II*) then reacts with an iron cyano complex, $[Fe(CN_6)]^{3-}$, to give Prussian Blue (Equation 3).

$$2\,[Fe(C_2O_4)_3]^{3-} \xrightarrow{hv} 2\,Fe^{2+} + 2\,CO_2 + 5\,C_2O_4^{2-} \qquad \text{Equation 2}$$

$$K^+ + Fe^{2+} + [Fe(CN)_6]^{3-} \rightarrow \underset{\text{Prussion blue}}{K[Fe(III)Fe(II)(CN)_6]} \qquad \text{Equation 3}$$

Through photochemistry, we are causing a redox reaction to occur. The key product of the reaction is the formation of Prussian Blue, which generates our image. The image is "fixed" by washing with water to remove any unreacted iron(*III*) oxalate and thus prevent further reactivity. To darken the image, you will submerge the film in hydrogen peroxide, which is a powerful oxidizing agent. This should oxidize the oxalate even further, thus providing a deeper blue color and greater overall contrast.

OVERVIEW

In today's lab you will explore the chemistry of cyanotype photography. To begin, you will devise a test plate to determine the optimal thickness of the photosensitizer solution and the required duration of UV exposure for your "film". By devising and testing these two variables, you will ensure the optimal clarity of your final image. Then, you will place an object of your choice on the prepared "film." The object will act as mask (also called a photoresist) to block the UV light and prevent the covered region from undergoing the chemical reaction. After ample exposure to UV light, the image will be developed by rinsing the film under water and by further oxidizing with hydrogen peroxide. The same procedure is then repeated using sunscreen as the photoresist.

PROCEDURE

Chemicals Used	Materials Used
Iron (III) oxalate sensitizer solution Hydrogen peroxide, H_2O_2 Sunscreen (high SPF)	Artist's water color paper (3) Black construction paper Paintbrush Tongs Overhead transparency Hair dryer Sun ☺ Bins for water and hydrogen peroxide Paper clips Items to be photographed (student provided)

Part A. Determining Exposure Time and Thickness of the Sensitizer Solution

The goal for this part of the experiment is to create a test plate to analyze the variables, exposure time and sensitizer film thickness, to determine which conditions are optimal for developing your image. Exposure time to form a clear image should take approximately 5 minutes on a sunny day. Depending on the day, however, the test range of exposure times can be as short as 1 minute and as long as 15 minutes. You will know the reaction time is sufficient when the exposed paper is dark. To determine the optimal thickness of sensitizer solution, you should test a light and a heavy covering.

Create your own test plate following your plans from PRE-LABORATORY EXERCISE #7. Take a piece of watercolor paper and in a room with no lights on (ambient light is okay) apply the sensitizer solution in the form of your test plate. Make sure the variables are labeled on your test plate. Then using a hair drier, thoroughly dry the paper. Cover your "film" with a notebook and carry it outside. Once outside, place your "film" in direct sunlight and expose the "film" according to your test plate protocol. After you have completed the test plate, cover the "film" again and bring the covered "film" inside.

Once inside, remove the covering and lay the film in a bin of water for ten minutes. Remove the film and rinse in the sink under cool running water. The image should be "fixed" by immersing the film in a bin of 3% hydrogen peroxide for 10 seconds, followed by a thorough rinsing with water. Allow the film to thoroughly dry on your desktop. When the test plate is thoroughly dry, determine the optimal thickness of sensitizer solution and exposure time.

Part B. Your image

Using a clean sheet of watercolor paper, repeat the procedure from Part A using the combination of thickness and exposure time that gave the best image in your test plate. Choose an item you wish to print (the photoresist) and clip it in place on the "film". Expose to the sun and repeat the developing process.

Part C. Sunscreen

Using a new sheet of watercolor paper, repeat the procedure a final time. Use a paintbrush to paint an image on a transparency using thickly applied sunscreen. Place the transparency on the prepared "film", cover the image, expose to the sun, and develop.

Cyanotype Photography as an Example of a Photochemical Redox Reaction

Name:	Lab Instructor:
Date:	Lab Section:

PRE-LABORATORY EXERCISES

1. Bring in an item you would like to make an image of. Some possibilities are leaves, paper cut outs, drawings using black markers on transparencies, lace, etc...

2. Define all underlined words in the BACKGROUND section.

3. Given the following reaction, label the oxidation state of each species involved. What species is oxidized? Reduced?

$$Mg\ (s) + 2HCl\ (aq) \rightarrow MgCl_2\ (aq) + H_2\ (g)$$

4. In today's experiment, why should the lights be off when you are making your cyanotype "film"? Would this be more of a problem if you were using silver bromide film? Explain.

OVER →

PRE-LABORATORY EXERCISES continued...

5. What is the wavelength range of UV light? Calculate the energy of the photons at the longest wavelength and the shortest wavelength in the UV region. Express your answer in both energy/photon and energy/mol of photons.

6. How do sunscreens work? What results to you expect in Part C of the experiment?

7. In the space below draw a copy of your planned test plate for Part A of the experiment.

Cyanotype Photography as an Example of a Photochemical Redox Reaction

Name:	Lab Instructor:
Date:	Lab Section:

RESULTS and POST-LABORATORY QUESTIONS

1. Attach a copy of your test plate and all the images that you made. Discuss the process of creating the test plate. What variables did you examine? What were the results?

2. Write a short essay explaining how the more traditional form of photography using silver bromide film works. You may need to consult your text-book or the internet as a source (include the name of any textbooks or the URLs and the date accessed). Your essay should include a description of the chemical processes that occur and the steps needed to make a final photograph.

A Titration for the Determination of Ions in Water: The Hard Truth

Kristen Spotz

OBJECTIVES

- Determine the ions responsible for hard water.
- Determine the relative hardness of tap water by titration.
- Explore commercial methods for water softening.
- Develop the laboratory skills required for accurate and precise quantitative measurements.

INTRODUCTION

You are strolling the aisles of the grocery store in search of the hand soaps when you spot two bar soaps, one specifically designed for hard water and the other for soft water. Would you know which brand to buy?

You receive your electric bill and it is significantly higher than normal. Your neighbor suggests investigating the hardness of your water. Would you understand the possible connection?

On the news you hear conflicting health reports concerning the benefits of having hard versus soft water. What would you do? Your first step to being a more informed consumer involves becoming more aware of what exactly is in your water that causes the condition known as water hardness.

BACKGROUND

Water and Water Hardness

The water we drink and use to cook, bathe and do our laundry is hardly pure. In fact tap water contains many "surprises", one of which is dissolved ions. In particular, two metal ions, which you will determine in this experiment, are largely responsible for the condition known as water hardness. One problem with hard water occurs when the hard water ions bind to soap to form insoluble precipitates which reduce the effectiveness of the soap. These solid precipitates cause soap scum or "bath tub ring" and wind up floating in water and later settling to form a sticky, gummy substance on the bottom of sinks, tubs, showers, and other water-using appliances.

Another problem with hard water occurs when hard water ions are heated causing a solid rocklike deposit of calcite crystals to form. These calcite crystals eventually cause a build-up of "scale" within plumbing as shown in the cross section of the pipe in Figure 1. Scale decreases the overall efficiency of water using appliances and leads to greater fuel consumption and higher utility bills. Hard water clogs your pipes, prevents soaps from sudsing and lowers the efficiency and the life-span of water-using appliances. An analysis of hard water gives an indication about the overall quality of the water supply.

Figure 1: Cross-section of a pipe clogged with scale

Analyzing Hard Water

In this experiment, the degree of hardness will be determined using an analytical method known as a titration. The titration will be performed using ethylenediaminetetraacetic acid, EDTA. EDTA is a weak polyprotic acid and can lose four hydrogens at high pH (basic solution) resulting in the ethylenediaminetetraacetate anion (Figure 2).

Figure 2: The ethylenediaminetetraacetate anion

In the EDTA molecule, four of the oxygen atoms, along with the two nitrogen atoms, can each grab onto a metal cation by donating their lone pair of electrons. Therefore, the EDTA molecule binds to a single metal cation in up to six places (like the fingers of your hand wrapping around a ball), which makes the resulting complex very strong. When EDTA is introduced into a given sample of water, the EDTA complexes with various cations, including any present hard water ions, in a 1:1 mole ratio (Equation 1 and Figure 3).

$$X^{n+}\ (aq) + EDTA^{4-}\ (aq) \rightarrow X(EDTA)^{(4-n)}\ (aq)$$

Equation 1: X^{n+} represents the hard water ions and $X(EDTA)^{(4-n)}$ represents the "complex ion".

We can determine the quantity of hard-water ions in solution by adding EDTA drop-wise until we have added enough to form a 1:1 complex with all the hard water ions in solution. If we keep track of how much EDTA we add, we can make a conclusion about the amount of hard water ions that must be present. This point, where we have added just enough EDTA, is called the endpoint or equivalence point of the titration. The equivalence point is defined as the quantity of titrant (the substance added) necessary to complex with the analyte (the substance being analyzed). The problem is how do we know when we have reached the endpoint and have added just enough EDTA?

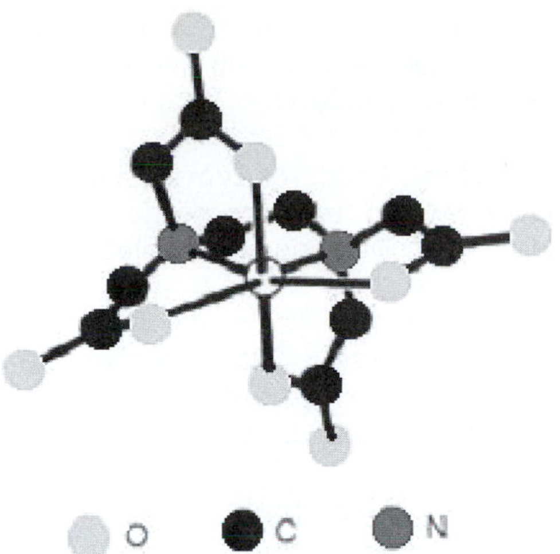

Figure 3: A three-dimensional model of the EDTA complex. The white circle in the center represents the hard water ion.

In order to detect the endpoint in this particular titration we will use the indicator, Eriochrome black T, along with a source of magnesium ions, $MgCl_2$. The magnesium is needed because the indicator, which is normally **blue** will form a **pink** complex in the presence of magnesium ions. As we begin our titration, the sample is **pink** due to the presence of indicator and magnesium ions (everything else in the solution, including the hard water ions, is colorless). As EDTA is introduced into the water sample, the hard water ions bind to the EDTA molecule in a 1:1 mole ratio. Once the EDTA is bound to all the hard water ions, the EDTA begins to pull the magnesium ions away from the indicator, In (Equation 2). The endpoint is noted when the solution turns **blue** due to the lack of magnesium ions that are available to bind with the indicator.

$$\underset{\text{Pink}}{\mathbf{MgIn\ (aq)}} + EDTA^{4-}\ (aq) \rightarrow Mg(EDTA)^{2-}\ (aq) + \underset{\text{Blue}}{\mathbf{In^{2-}\ (aq)}}$$

Equation 2: Occurs after all of the hard water ions in the sample are first complexed with EDTA.

PROCEDURE

Part A: Metal Ions Responsible for Hard Water

You will determine the ions responsible for hard water by determining which metal ions reduce the amount of sudsing.

Chemicals Used	Materials Used
Magnesium nitrate	Test tubes (5)
Sodium nitrate	Rubber stoppers (5)
Calcium nitrate	Metal spatula
Potassium nitrate	Test tube rack
De-ionized water	Analytical balance
Liquid Ivory™ soap	Disposable dropper

1. Label the test tubes 1-5. Each of the first four test tubes will contain about 0.5 grams of one of the following compounds: sodium nitrate, calcium nitrate, magnesium nitrate, or potassium nitrate. Record the contents of each test tube.

2. Half-fill each of the five test tubes with de-ionized water. Test tube #5 has only de-ionized water and will serve as a control. Stopper each test tube and shake until each solute is dissolved.

3. Add three drops of Ivory™ soap to each test tube. Shake vigorously for 5 seconds and allow the contents to settle. Sketch your observations of the level of sudsing for each test tube. Dispose of all solutions down the sink.

Part B: Titration of Hard Water

You will perform titrations to determine the hardness of a sample of local water. You will begin your analysis by titrating a blank.

Chemicals Used	Materials Used
EDTA stock solution	Ring stand and buret clamp
Buffer solution (pH 10)	250-mL Erlenmeyer flasks (4)
Indicator (Eriochrome black T)	10-mL Graduated cylinders
Magnesium chloride solution	100-mL, 250-mL Beakers
De-ionized water	50-mL Buret
Various water samples	Funnel
	Hot plate
	1-mL, 50-mL Volumetric pipets and pipet bulb

CAUTION: Be careful, hot plates can stay warm for a long time after they are turned off.

1. Prepare a sample of tap water for study by first allowing the water to run about two minutes to flush the pipes. Take about 250 mL of the tap water and set aside for study as described below.

2. Prepare a 50-mL buret. Begin by running about 50 mL of de-ionized water through the buret and then rinse the buret with about 5.0 mL of EDTA solution. Use a 100-mL beaker to collect the waste.

3. Mount the buret in the ring stand and, using a funnel, fill the buret with stock solution of EDTA. Record the concentration of EDTA solution and the initial buret volume.

4. Prepare the blank solution to be titrated. Begin by pipeting 50.00 mL of de-ionized water into a 250-mL Erlenmeyer flask. Record the color after each of the following additions (placing a piece of white paper under the flask will help you accurately see the color). Add 5.00 mL of buffer and 2-3 drops of the indicator, Eriochrome black T. Pipet 1.0 mL of the $MgCl_2$ solution to the blank while swirling the flask. Heat the solution to 50-60°C. Remove the flash from the hot plate. The warm solution is now ready to titrate.

5. Titrate the solution in the flask with EDTA. The endpoint comes very soon with the blank titration so adjust the buret to dispense the EDTA drop-wise. Swirl the flask to mix the solution during the titration. The solution should turn from dark pink to dark purple to dark blue. When the last tinge of purple disappears, stop and allow the solution to sit for a few moments because the indicator is slow to react. Set this solution aside and save as a reference endpoint for the other titrations. Record the volume of EDTA (± 0.02 mL) needed to titrate the blank. This volume will be subtracted out in following titrations to account for the amount of magnesium ions added.

6. Prepare a sample of tap water for titrating. Follow the directions used in step 4 above, but substitute 50.00 mL of tap water for the 50.00 mL of de-ionized water used previously.

7. Refill the buret with EDTA, record the initial volume and titrate the solution of tap water following the directions in step 5 above. The endpoint will take longer to reach than with the blank.

8. Repeat steps 6 and 7 until you have two trials with volumes of EDTA (corrected for the blank) within 2% of each other. Error is calculated using the following formula:

$$\text{Error} = \left(\frac{\text{Trial I} - \text{Trial II}}{\text{Trial I}}\right) \times 100 \leq 2\%$$

9. All waste can be poured down the sink with copious amounts of water.

Part C: Water Softening

You will evaluate the effectiveness of a commercial water softener (Calgon™).

Chemicals Used	Materials Used
Calgon™ Tap water and de-ionized water	Well plate with 24 wells Plastic pipets and Glass stir rods

Instructions

1. Using a 24-well plate, half-fill wells A1 and A2 with de-ionized water and half-fill wells A3 and A4 with tap water. Add one drop of Calgon™ to wells A2 and A4 and stir.

2. To all of the wells (A1 - A4), add one drop of buffer and one drop of indicator. Note and record the color changes. All waste can be poured down the sink.

A Titration for the Determination of Ions in Water: The Hard Truth

Name:	Lab Instructor:
Date:	Lab Section:

PRE-LABORATORY EXERCISES

1. Read the section on titration in your textbook. In your own words, rewrite a brief description of what is involved in a titration.

2. Do you expect that de-ionized water would be hard or soft? Why?

3. Thanks to King Henry VIII, a common way of reporting water hardness is in grains per gallon where 1 grain=64.0 mg. If natural water averages a relative hardness of 6.00 grains per gallon convert this to mg/liter.

4. One way of combating hard water is to use soaps and detergents with added phosphates. Phosphates bind with the hard water ions to form insoluble precipitates, instead of the ions binding with the soap. Make a list, using the solubility rules in your text, of 6 insoluble ionic compounds that contain the phosphate ion.

OVER →

PRE-LABORATORY EXERCISES continued...

5. Explain the purpose of everything you added to make your blank in Part B, step 4. What is the purpose of titrating the blank?

A Titration for the Determination of Ions in Water: The Hard Truth

Name:	Lab Instructor:
Date:	Lab Section:

RESULTS and POST-LABORATORY QUESTIONS

Part A: Metal Ions Responsible for Hard Water

Indicate the ion present and sketch the relative level of water and suds for each test tube.

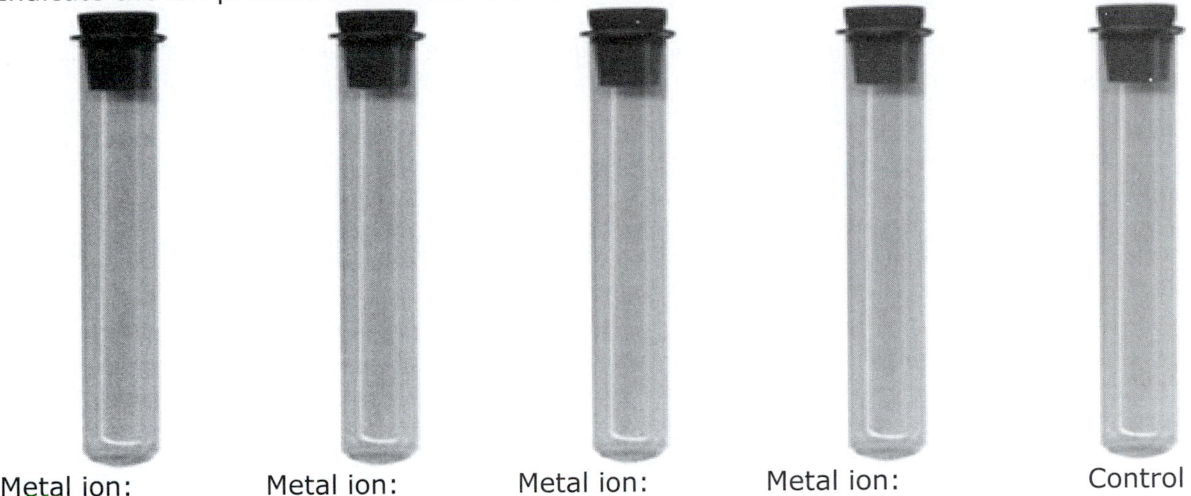

Metal ion: Metal ion: Metal ion: Metal ion: Control

Based on the sudsing test in Part A, which two metal ions are likely responsible for water hardness? Explain.

Part B: Titration of Hard Water

Volume of EDTA used to titrate blank

Tap water:	Trial #1	Trial #2	Trial #3	Trial #4
Initial buret reading				
Final buret reading				
Total volume of EDTA				
Vol. corrected for blank				

Put a star (*) next to the two trials that are within 2%.

Average corrected vol. of EDTA (use selected trials, *)	
Concentration of EDTA (from stock bottle)	
Number of moles of EDTA needed to titrate sample	
Number of moles of hard water ions present in sample	
Volume of hard water sample analyzed	
Concentration (mol/L) of hard water ions in water sample	
Class average: Concentration of hard water ions	

OVER →

RESULTS and POST-LABORATORY QUESTIONS continued...

Part B: Titration of Hard Water continued

1. What color was the blank solution after you added 5 mL of buffer and seven drops of indicator? What species were causing the color?

   What color appeared after the magnesium chloride was added? What species were causing the color?

   What color appeared after titrating with EDTA? What species were causing the color?

2. Research indicates that students who write about and talk about science in informal settings often exhibit a deeper level of content mastery. With this in mind, we ask you to write a hypothetical letter to a friend or relative explaining hard/soft water along with the consequences and drawbacks of hard water. Based on your knowledge of hard/soft water, would you recommend they soften their water and why? Include specific information about the actual hardness of your water (see the table below). You can be specific about what water in your house you would choose to soften. In addition, explain a sample test that they can perform to determine if their water is hard. When writing your letter, it is important to know that there are conflicting reports concerning hard/soft water. For example, softening water (removing the hard water ions) increases your risk of heart disease because sodium ions are "exchanged" for the hard water ions in the process. The U.S. Department of Agriculture's Human Nutrition Research Center has stated, "The health benefits of drinking hard water far outweigh the minor inconveniences." However, having hard water increases children's and infants' risk of developing eczema, a skin condition characterized by an itchy rash. This may be due to the increased amount of soap needed in hard water or to the actual minerals in the water.

Hardness Description	Hardness (mol/L)
Soft	$0 - 6.0 \times 10^{-4}$
Moderately Hard	$6.0 \times 10^{-4} - 1.2 \times 10^{-3}$
Hard	$1.2 \times 10^{-3} - 1.8 \times 10^{-3}$
Very Hard	$> 1.8 \times 10^{-3}$

Name:	Lab Instructor:
Date:	Lab Section:

RESULTS and POST-LABORATORY QUESTIONS continued...

Part C: The Effect of Commercial Water Softeners on Hard Water

Well	Contents	**Color** (after indicator and buffer)
A1		
A2		
A3		
A4		

Explain the significance of each of the four wells. Why was each well necessary to see if the water softener was working?

Did the commercial water softener appear to help soften the water? Explain.

Spectrophotometric Analysis: Phosphates in Water

Kristen Spotz

OBJECTIVES

- Practice calculating and performing dilutions of solutions.
- Determine the concentration of phosphate in a water sample by spectrophotometric analysis.
- Construct and utilize a calibration curve.
- Explore the dynamics of working with a larger group of students.

INTRODUCTION

Imagine a time when the lakes and rivers are no longer safe for swimming or boating, or when the ocean is no longer a source of food. Coastal zones and estuaries, some of the most productive ecosystems in the world, are in danger. The problem of eutrophication, is affecting the water supply of towns across the nation making the water unsafe for consumption and hazardous to the wildlife that depend upon it.

The source of the eutrophication problem is an excessive input of nutrients into rivers, lakes and the seas because of the extensive use of fertilizers, the combustion of fossil fuels and waste from animal feedlots. This excessive nutrient input stimulates the growth of algae and bacteria, robbing the water of precious oxygen. The resulting algal blooms, red tides and deterioration of sea grass makes the waters uninhabitable for most fish and coastal wildlife.

What role will you play as a future scientist or citizen in ensuring the protection of our valuable water resource?

BACKGROUND

Phosphates are one of the major groups of contaminates affecting our nation's water supply. Phosphates are found in the environment, not only in the form you have seen in your chemistry book (PO_4^{3-}), but also as polyphosphates (such as $P_2O_7^{4-}$ or $P_3O_{10}^{5-}$) or as organic phosphates which are eventually hydrolyzed to form PO_4^{3-}. The primary means by which humans introduce phosphates into the environment is through the use of fertilizers and detergents. In particular, tripolyphosphates ($P_3O_{10}^{5-}$) have been used in soaps and detergents to combat the problem of hard water. Phosphates are also a major component of fertilizers, because phosphorus is a necessary plant nutrient and is crucial for seed formation, root development, and crop maturation. These phosphates eventually enter the water supply leaving lakes, rivers, and seas with an abnormally high phosphate concentration.

Spectrophotometric Analysis and the Determination of Phosphate

Spectrophotometric analysis relies on the fact that the amount of light absorbed by a sample shows a linear dependence upon the concentration of the compound present in the solution. You have probably seen this phenomenon for yourself before. Just hold up two glasses of juice made from powdered concentrate; one made with three scoops and one made with one scoop. The more concentrated drink absorbs more light and is darker. The problem with using spectrophotometric analysis in our case is that phosphates are colorless and therefore do not absorb light in the visible portion of the electromagnetic spectrum. However, due to the reactive nature of phosphates, one can easily color them using an ammonium vanadomolybdate reagent. This reagent includes ammonium metavanadate (NH_4VO_3) and molybdate (MoO_4^{2-}) and reacts with the phosphate to form a yellow compound (called "heteropoly acid" from here on). The formula of the yellow compound is uncertain but thought to be $(NH_4)_3PO_4 \bullet NH_4VO_3 \bullet 16MoO_3$. The brightness of the resulting yellow solution is directly proportional to the concentration of phosphate in the water.

Scientists use an instrument called a spectrometer to quantitatively determine the amount of light absorbed by a solution. The primary inner parts of a typical spectrometer are illustrated in Figure 1. The spectrometer has a light source that emits light which is focused with a small slit. The wavelength of interest is then selected using the monochrometer ("mono" meaning one and "chromate" meaning color) and an additional slit. The selected light then reaches the sample and depending on how the photons interact with the compound of interest, the light is either absorbed or passes straight through. By comparing the amount of light entering the sample (P_o) with the amount of light reaching the detector (P), the spectrometer is able to tell how much light is absorbed.

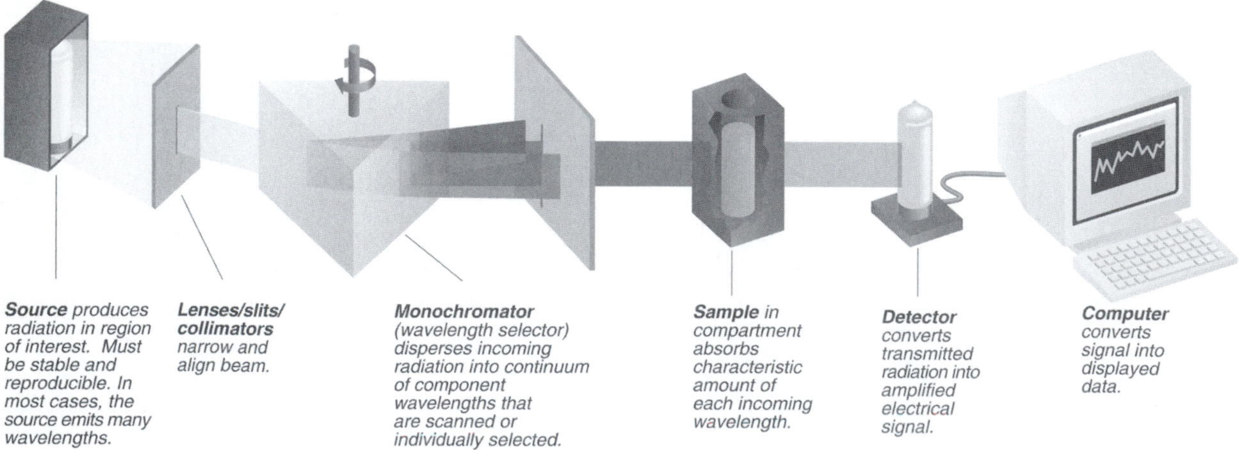

Figure 1. A schematic diagram illustrating the interior components of a typical spectrometer.

Scientists quantify the amount of light passing through the sample in terms of percent transmittance (%T). Percent transmittance is calculated as the fraction of original light that passes through a sample (Equation 1).

$$\%T = \left(\frac{P}{P_o}\right) \times 100 \qquad \text{Equation 1}$$

Equation 2 shows how percent transmittance (%T) can easily be converted into a quantity known as absorbance (A). Though most spectrophotometers give readings in terms of both %T and A, measurements should be made in %T and mathematically converted to A because %T can be determined more accurately.

$$A = -\log\left(\frac{\%T}{100}\right) \qquad \text{Equation 2}$$

The absorbance of a sample is important because of the previously mentioned linear relationship between absorbance the concentration of the sample. This relationship is known as Beer's law (Equation 3).

$$A = \varepsilon bc \qquad \text{Equation 3}$$

The amount of the light that is absorbed depends on several variables:

- "A" is the absorbance of the sample, which in this experiment is due to the interaction of phosphate, in the form of heteropoly acid, with the photons of light. Although the compound being studied may, in general, absorb light over a fairly broad range of wavelengths, there is only one region where the light is absorbed most strongly. This wavelength is known as λ_{max} (pronounced "lambda max"). The absorbance of the sample should be measured at this wavelength.

- "ε" is the molar absorptivity. The molar absorptivity is a constant representing the efficiency by which the substance absorbs light. The greater the value of "ε" the more strongly the substance absorbs light resulting in a more intense color.

- "b" represents the solution path length. It is the distance that the light must travel through the sample and is measured as the width of the sample holder (also called a cuvette). "b" is a constant for each experiment (typically 1 cm).

- "c" represents the molar concentration of the absorbing species in the sample.

One can easily determine the unknown concentration of a sample from Equation 3 after measuring the absorbance of the sample and using the molar absorptivity of the compound and the path length of the cuvette. If the molar absorptivity of the compound is not known, the concentration of an unknown can still be found by constructing a calibration curve.

The Calibration Curve

A calibration curve allows scientists to determine the unknown concentration of a known species. According to Beer's law, as long as we account for a blank solution in our studies, a plot of absorbance versus concentration gives a straight line with slope = "εb" and a y-intercept = 0. For example, the calibration curve in Figure 3 is used to determine the concentration of an unknown solution of iron. The graph is constructed from six points that are made from a stock solution of iron having a known concentration. The experimentally measured absorbance of each of the six solutions is then plotted as a function of concentration and a line of best fit is drawn through the points. As expected, the absorbance of the sample increases linearly as the concentration increases. The absorbance value of 0.357 was then measured for the unknown iron solution of interest. To relate the absorbance to the unknown concentration we can either use the equation of the line of best fit or we can extrapolate from the graph (as shown in Figure 2). This absorbance value was found to correspond to a concentration of 3.59 M of iron in the unknown sample.

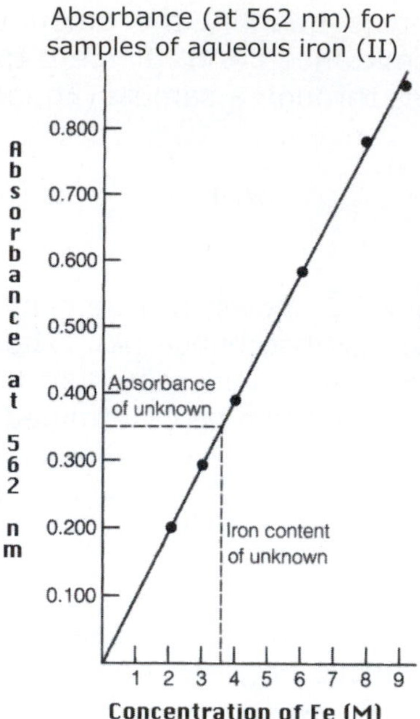

Figure 2. A sample calibration curve used to determine the concentration of an unknown sample of ion (II)

The calibration curve in Figure 2 is an example of a successfully constructed graph. The title is labeled above the graph with both the axes clearly labeled using the independent (x-axis) and the dependent (y-axis) variables in the experiment. After plotting each of the data points on the graph, a line of best fit is drawn. Although, the points do not have to fall directly on the line, a good agreement is expected and needed for accurate determination of the concentration of your unknown.

OVERVIEW

In this experiment, students will work in groups to first prepare a series of six standard solutions of known phosphate concentration by dilution of a stock solution. Using λ_{max} of 400nm, the absorbance of the five standard phosphate solutions will be measured and used to construct a calibration curve. The absorbance of a sample of unknown phosphate concentration will then be determined. The calibration curve will be used to relate the absorbance to the unknown concentration of phosphate in the sample.

PROCEDURE

Chemicals Used	Materials Used
Phosphate stock solution (1.00 X 10^{-3} M) Ammonium vanadomolybdate solution Various water samples of unknown phosphate concentration	Spectrophotometer 25-mL Volumetric flask 1, 5, 10 and 25-mL Pipets and pipet bulb 10-mL Cuvette (1 per group of students) 100-mL Beakers

Part A: Organizing your group

1. Students will work in groups of 2-3 to construct a single calibration curve consisting of 6 data points having phosphate concentrations in the range 4.00 x 10^{-5} M to 4.00 x 10^{-4} M. Each student will be responsible for making at least one of the solutions and measuring the absorbance for at least one data point. Show your instructor your calculations for making your 25-mL standard solutions from the 1.00 X 10^{-3} M phosphate stock solution before your group goes on to Part B. Remember, solutions must be made using only the available volumetric flask and pipets.

Part B: Adjusting the Spectrophotometer

2. Turn on the spectrometer (Figure 3) by rotating the power control clockwise. Allow the spectrophotometer to warm-up for five minutes before using.

3. Adjust the wavelength to 400 nm. With no sample in the spectrometer, turn the zero adjust so the meter reads 0% T. Each member of the group should verify all readings.

4. Prepare the blank by pipetting 10 mL of de-ionized water and 5 mL of ammonium vanadomolybdate into a beaker.

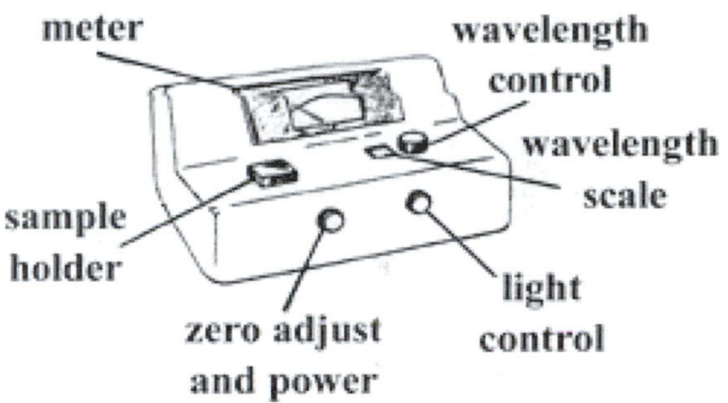

Figure 3. A typical spectrometer

5. Always rinse the cuvette with a few mLs of solution whenever you are using a new solution. Discard the rinsing solution according to your instructor's directions. Three-quarters fill the rinsed cuvette with the blank solution. Insert the cuvette into the sample holder of the spectrometer and adjust the light-control knob so 100% transmittance is read. Your instrument is now zeroed.

Part C: Preparation of Standard Solutions

6. Based on your calculations from Part A, pipet the appropriate volume of the 1.00 x 10^{-3} M phosphate stock solution into a 25-mL volumetric flask. Dilute the stock solution by filling the volumetric flask until the meniscus reaches the mark (Figure 4).

7. Pipet 10.0 mL of the phosphate solution you made in step 6 and 5.00 mL of the ammonium vanadomolybdate stock solution into a small, labeled beaker.

8. Repeat steps 6 and 7 for each of the six standard solutions.

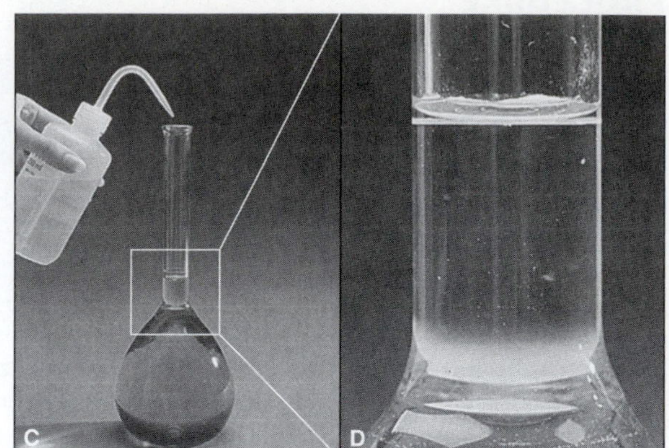

Figure 4: Proper dilution of a solution by filling a volumetric flask to the line

Part D: Making the Calibration Curve Using the Standard Solutions.

9. Rinse the same cuvette you used for your blank with about 1 mL of your standard solution (from step 7). Three-quarters fill the rinsed cuvette with the sample solution. Insert the cuvette into the spectrometer. Measure and record the percent transmittance. All data points for a given curve must be measured with the same cuvette. All phosphate solutions should be discarded according to your instructor's directions.

10. Repeat step 9 for the remaining phosphate solutions that your group made. Before using any glassware with each new solution, the glassware must be rinsed with de-ionized water and about 1 mL of the new solution.

Part E: Determination of Unknown Concentration

11. Pipet 10.0 mL of the unknown and 5.00 mL of the ammonium vanadomolybdate solution into a beaker. Half-fill the rinsed cuvette with the unknown solution. Use the spectrometer to measure the percent transmittance. Record your results.

Before you leave: Make sure everyone in your group has recorded the concentration and the %T for each of the various phosphate solutions.

Spectrophotometric Analysis: Phosphates in Water

Name:	Lab Instructor:
Date:	Lab Section:

PRE-LABORATORY EXCERCISES

1. Define the underlined words in the BACKGROUND section.

2. In your own words, summarize the purpose of a calibration curve.

3. To prepare yourself for performing the dilutions required in this laboratory experiment, read the section on dilutions in your textbook. What volume of 1.00×10^{-3} M phosphate stock solution is required to make 25.0 mL of a 4.00×10^{-5} M solution?

4. Using the spectrophotometer, a sample was analyzed and found to have a percent transmittance of 85%.
   a) What percent of light was actually absorbed by the sample?

   b) Calculate the absorbance (A) of the sample.

Spectrophotometric Analysis: Phosphates in Water

Name:	Lab Instructor:
Date:	Lab Section:

RESULTS and POST-LABORATORY QUESTIONS

1. Attach a copy of your data table from today's experiment. Your table should include the concentration of phosphate in each standard solution, the measured %T and your calculated absorbance.

2. Attach a copy of your calibration curve. What is the equation of the best-fit line?

3. Determine the concentration of phosphate in your unknown solution by extrapolation of the calibration curve (refer back to Figure 2) and by using the equation for the line of best fit. The extrapolation should be shown on your attached calibration curve. The calculation using the line of best fit should be shown below.

4. The U.S. Public Health Service has set the maximum value of phosphate in the drinking water at 0.30 mg phosphate/liter. Did your unknown water sample violate this standard? Show your work.

Discovering the Gas Laws

Brian Polk

OBJECTIVES

- Explore the relationships between the pressure, volume, quantity and temperature of a gas.
- Replicate the gas law experiments of Robert Boyle, Jacques Charles, and Amadeo Avagadro.

INTRODUCTION

Take a deep breath, hold it for a moment and then exhale. Have you ever wondered why air enters and leaves your body when you breathe? You are able to take air into your lungs because as you inhale, your diaphragm contracts which expands the volume of your lungs. The increased volume results in a decrease in the pressure of gases inside your lungs. The external pressure (P_{atm}) is now greater than the pressure in your lungs and the atmosphere pushes air into your lungs (Figure 1). The reverse process happens when you exhale.

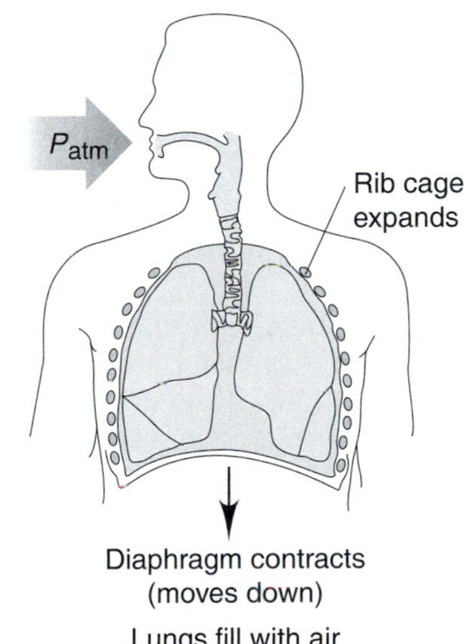

Figure 1. Taking a breath is a real world application of Boyle's Law, which relates the pressure and volume of a gas.

The act of breathing serves to illustrate the general relationship between the pressure and volume of a gas. As a practical example of why it is important to understand this relationship, consider the fact that submarine crew members are instructed that in the event of an emergency underwater evacuation they are not to hold their breath as they surface, which is the common response. Rather they are told to slowly exhale as they come to the surface. Can you explain why?

BACKGROUND

Properties of Gases

As scientists, we are interested in studying the physical properties of gases. For example, the gases you breathe are generally colorless and odorless. Scientists also describe gases as having a low density, low viscosity and high miscibility. As a result, gases are typically found in nature as solutions.

In order to understand the origin of some of these properties, it is useful to have a conceptual understanding of what occurs in a gas on the atomic level. Let's begin by comparing the three phases of matter (Figure 2). The particles (the atoms, molecules or ions) in a solid are held close together because they do not have sufficient kinetic energy to overcome the interactions between the particles. The result is that solids have a definite shape and volume. Liquids also have definite volume, but because the particles in a liquid have more mobility than in a solid, liquids have an indefinite shape. By comparison gases are mostly empty space and have both an indefinite shape and volume. The particles making up a gas are far apart from each other and move very quickly. The average oxygen molecule in the air around you is moving at roughly 1000 miles/hour! As a result, gases flow freely, are compressible, and diffuse through space and through other gases to fill the available space.

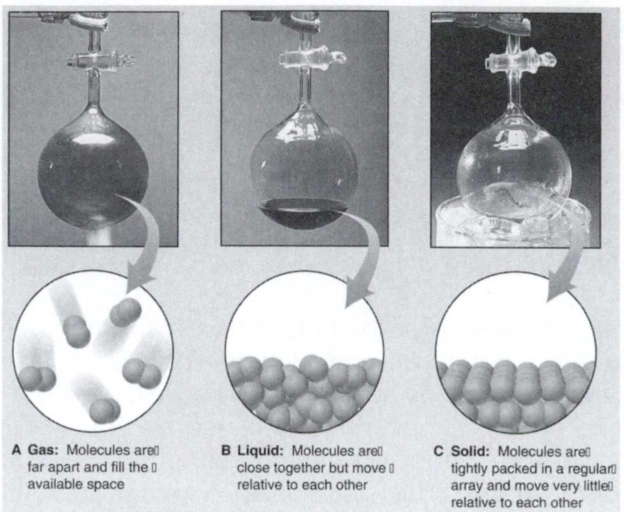

Figure 2. Atomic scale representations of the gaseous (left), liquid (center), and solid phases (right) of bromine.

Measuring Pressure, Volume, Temperature and Quantity of Gases

In this experiment we will be reproducing the results of several great scientists who studied the relationship between the following properties of gases:

Pressure (P): In some ways, pressure is the most interesting of these related properties. You might not think that something that is mostly empty space and that you can't see would be able to exert a pressure. Yet, the force we experience on the surface of the Earth, due to the atmosphere, is about 14.7 lb/in^2 (1 atm). How do the particles of a gas create pressure? Gas particles are in constant, random, straight-line motion, except when they collide with the container walls or with each other. Imagine the Lottery drawing where the balls bounce around and never stop moving. As gas particles collide with the surface of the container they exert a measurable pressure (Figure 3).

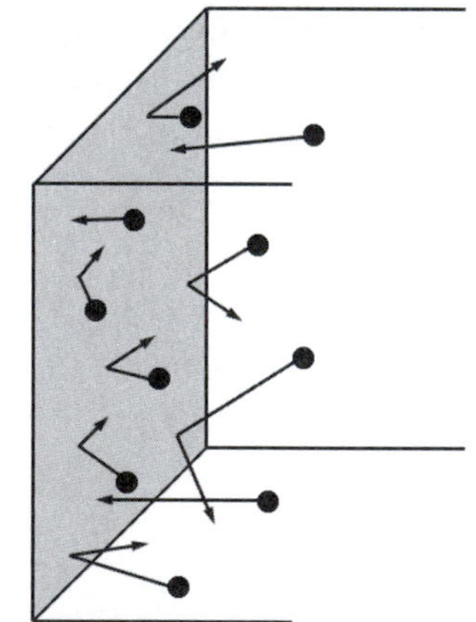

Figure 3. Gas particles exert pressure through constant collisions

Volume (V): Because gases fill their container, scientists typically quantify the volume of gas as equal to the volume of the container that contains the gas. For example, a 2-liter bottle, without the liquid, would contain 2 liters of gas.

Temperature (T): Temperature is the measure of how hot or cold a substance is relative to another substance. The temperature of a gas is usually measured with a thermometer and is expressed in terms of the Kelvin scale. Temperature can also be understood as a measure of the energy of molecular motion. The average gas particle in a sample of hot air moves faster than in cold air. As a direct result, different gases at the same temperature have the same average kinetic energy.

Quantity (n): The individual particles in a gas are obviously too small to count. For this reason, it is convenient to quantify the amount of gas present in terms of the mole.

Boyle's law

The first scientist to explore the relationship between pressure and volume was Robert Boyle in 1662. He determined that volume and pressure of a gas are inversely proportional:

$$V \propto \frac{1}{P} \qquad \text{Equation 1}$$

or,

$$V = \frac{\text{constant}}{P} \qquad \text{Equation 2}$$

If we are dealing with changes in volume (or pressure) and we want to calculate the resulting change in the other variable, it is common to express Boyle's law in the form:

$$\frac{V_1}{V_2} = \frac{P_2}{P_1} \qquad \text{Equation 3}$$

In this form, the volume (V_1) is still inversely proportional to the corresponding pressure (P_1).

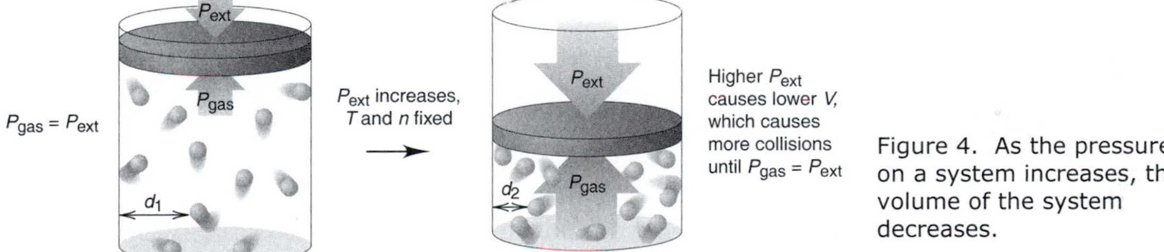

Figure 4. As the pressure on a system increases, the volume of the system decreases.

Suppose that the container on the left in Figure 4 has a volume of 16 L (V_1) and an initial internal and external pressure of 1 atm (P_1). When an external pressure of 2 atm (P_2) is applied to the system, at constant n and T, the volume of the container decreases to 8 L (V_2) as calculated using Equation 3. As the external pressure increases, the distance between the gas particles decreases and the volume decreases. The pressure of the gas increases because the frequency of collisions between particles and with the container increases with the smaller space between particles. This relationship should be intuitive. As you increase the external pressure by squeezing a capped, empty plastic soda bottle, what happens to the volume of the gas inside the bottle?

Charles's Law

On June 5th, 1783, Joseph and Etienne Montgolfier used fire and hot air to inflate a spherical balloon about 30 feet in diameter. The balloon traveled 1.5 miles before it came back to earth. News of this remarkable achievement spread throughout France, and Jacques Charles immediately tried to duplicate this feat. As a result of his work with balloons, Jacques

Charles noticed that the volume of a gas is directly proportional to its absolute temperature (Kelvin scale) as expressed in Equations 4, 5 and 6:

$$V \propto T \qquad \text{Equation 4}$$

or,

$$V = \text{constant} \times T \qquad \text{Equation 5}$$

Again, when performing calculations, Charles's law is commonly written as shown in Equation 6. The temperatures used in Equation 6 must be in the Kelvin scale.

$$\frac{V_1}{V_2} = \frac{T_1}{T_2} \qquad \text{Equation 6}$$

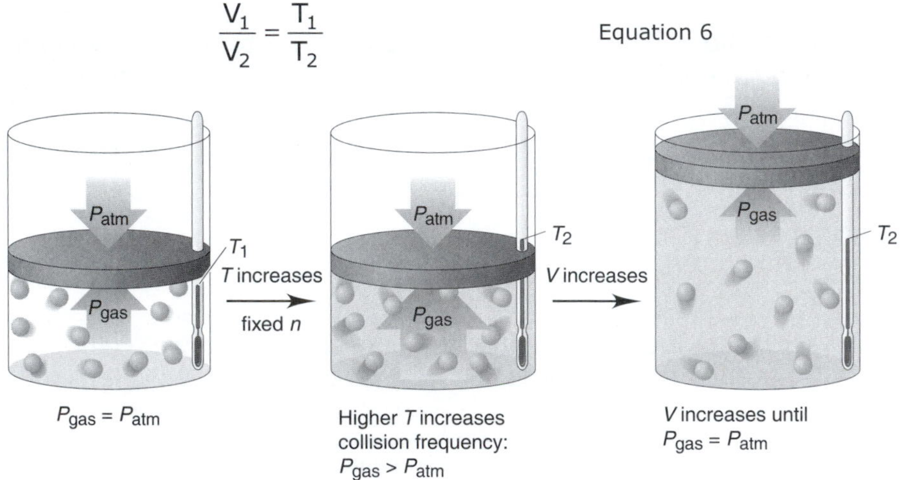

Figure 5. As the temperature of a system increases, so does the volume

Figure 5 shows that as the temperature of a system increases, so does the pressure due to the increased number and kinetic energy of the collisions between the gas particles and the walls of the container. If the container is designed to allow for constant pressure (that is, the container can expand or contract), then the volume of the container will increase until the pressure of the gas is equal to the pressure of the atmosphere. Again a practical example serves to illustrate the intuitive nature of this relationship. In areas of the country with extremes in annual weather, there is a noted fluctuation in the volume of car tires.

Avogadro's Law

The final piece to the puzzle was provided by Amadeo Avogadro. Avogadro's law can be stated in several different ways but essentially says that the volume of a gas at constant temperature and pressure is directly proportional to the number of moles of the gas (Equations 7 and 8). As the number of particles of gas in a container increases, so does the number of collisions with the walls. The volume of the container will expand until this increased number of collisions exerts the same pressure per unit of wall as before the addition.

$$V \propto n \qquad \text{Equation 7}$$

or,

$$V = \text{constant} \times n \qquad \text{Equation 8}$$

It is significant to note that Avogadro's law is stated independently of the identity of the particular gas. This implies that equal volumes of any ideal gas (at the same pressure and temperature) will have the same number of particles (or moles).

OVERVIEW

In this experiment you will perform three separate experiments to analyze the effect of pressure, number of moles and temperature on the volume of a gas.

PROCEDURE: For this experiment, you should sketch a rough graph of your data as you go along to see if you need to redo any of the measurements.

Part A. Boyle's law

Materials Used
35-mL Syringe sealed with heat or epoxy Bathroom scale Syringe cap or small board with a hole to accommodate the syringe head Ruler

After zeroing the scale, place the syringe cap or the small board in the center of the scale. Place the sealed end of the syringe vertically into the cap (or the hole in the board). With the palm of your hand, push down on the plunger of the syringe. As you decrease the volume on the syringe stop at 5-mL increments and record the "weight" shown on the bathroom scale. The pressure you are recording is in pounds (from the scale reading) per square inch (from the surface area of the cap or the board).

Part B. Charles's law

Materials Used	
Hot plate Ring stand and clamp 25-mL Lubricated syringe	Thermometer 125-mL Erlenmeyer flask 600-mL Beaker Rubber stopper with one hole

Cap an Erlenmeyer flask with the rubber stopper. Insert a 25-mL syringe into the hole, making sure it is snug and that the plunger on the syringe is generously lubricated. Set up a thermometer, flask, water bath and hot plate as shown in Figure 6. Slowly heat the system to 100°C while stirring the water bath. Record the volume of the syringe every 5°C.

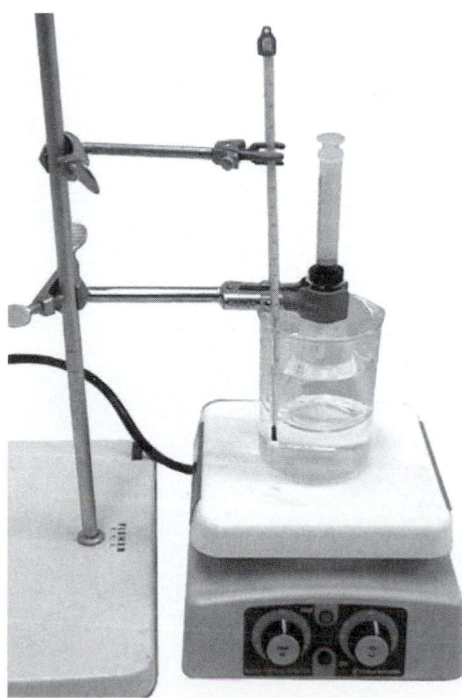

Figure 6. Set-up for Charles' law

Part C. Avogadro's law

Chemicals Used	Materials Used
Methanol (CH_3OH) Ethanol (C_2H_5OH)	Hot plate Thermometer 50-mL Erlenmeyer flask 125-mL Erlenmeyer flask 10-mL Graduated cylinder Aluminum foil (small piece) Pin Tongs Analytical balance

CAUTION: Ethanol and methanol are both flammable and should be handled with great care. Do not use any open flames during this lab.

Your instructor will assign you to work with either methanol or ethanol. Weigh a 50-mL flask. Pour about 10 mL of ethanol or methanol into the flask. Cover the top of the flask with a piece of aluminum foil that has a very small pin-hole in it. Heat the flask on a hot plate until all the liquid has evaporated. Carefully remove the hot flask and allow the sample to slowly cool to room temperature. Reweigh the flask and condensed gas. Repeat the experiment with a 125-mL flask. Calculate the number of moles of methanol or ethanol present in each flask and write your answers on the board with the class data.

Discovering the Gas Laws

Name:	Lab Instructor:
Date:	Lab Section:

PRE-LABORATORY EXERCISES

1. Define the <u>underlined</u> words in the BACKGROUND section.

2. If the initial volume of a container is 2.50 L and the temperature is 150.0 °C, what is the volume if the temperature is decreased to 60.0 °C?

3. The following questions refer to Avogadro's law. Read the appropriate sections in your text before answering.

   a) Look at Equations 3 and 6. What is the corresponding equation for Avogadro's law?

   b) Look at Figures 4 and 5. Draw a corresponding picture for Avogadro's law.

   c) Give a practical example illustrating Avogadro's law.

   d) Use Avogadro's law to explain why a 1 liter balloon filled with helium is less dense than a 1 liter balloon filled with nitrogen.

OVER →

PRE-LABORATORY EXERCISES continued...

4. For Part C, should the aluminum still be on the flask when you reweigh the flask and condensed gas? Explain.

5. What is an "iron-lung"? Briefly explain how it works in terms of the gas laws. Include a reference for your source. If you used the Internet, include the URL and the date you last accessed it.

Discovering the Gas Laws

Name:	Lab Instructor:
Date:	Lab Section:

RESULTS and POST-LABORATORY QUESTIONS

Part A. Boyle's law

Attach a table of your data. You should include columns for P (scale only, lb/in^2), P (scale only, atm), P (total, atm), 1/P (1/atm) and V (mL). Be sure to account for the total pressure on the gas. That is, account for both your pressing on the syringe (from the scale reading) and the atmosphere pushing down (from barometer and information in BACKGROUND).

Attach a plot of V (mL) versus P (atm) and a plot of V versus 1/P for your data in Part A.

Which of the two plots gives a linear correlation? Does this make sense? Explain.

What is the slope of the line in the linear plot?

Use the slope of the line to predict the pressure of the gas when the volume is 0.50 mL.

Part B. Charles's law

Attach a table of your data. You should include columns for T (°C) and V (syringe, mL) and V (total: syringe + flask, mL).

Attach a plot of your data for Part B as V (total) versus T (°C). Discard data from before the syringe moves and after the maximum volume of the syringe is reached. The remaining data should be linear.

What is the equation for the straight line?

Predict the temperature of the gas corresponding to a volume of 0 L. What is the significance of this result?

OVER →

RESULTS and POST-LABORATORY QUESTIONS continued...

Part C. Avogadro's law

50-mL flask	methanol / ethanol (Circle the gas you studied)	methanol / ethanol (Circle the other gas)
Weight of empty flask		
Weight of flask + condensed gas		
Weight of gas		
Molar mass of gas		
Moles of gas		
Class data: ave. moles of gas		

125-mLflask	methanol / ethanol (Circle the gas you studied)	methanol / ethanol (Circle the other gas)
Weight of empty flask		
Weight of flask + condensed gas		
Weight of gas		
Molar mass of gas		
Moles of gas		
Class data: ave. moles of gas		

Examine the average class data from Part C. What two major conclusions can you draw and how do they relate to Avogadro's law?

Use your results to predict the mass of gaseous methanol that would occupy a 175-mL flask.

Gas Stoichiometry: The Automobile Airbag

Joel Kelner

OBJECTIVE

- Design a model airbag as an example of a gas stoichiometry problem.
- Develop a laboratory protocol to solve a problem.
- Write a laboratory report.

INTRODUCTION

Since the late 1980's, thousands of lives have been saved by the collision protection provided by airbags (Figure 1). The death rate for drivers utilizing both seat belts and airbags is about 30% lower than the death rate among drivers using only seat belts.

Within 40 milliseconds after the initial automobile impact, the airbag is fully inflated, providing the crucial cushion between a possible life or death situation for the driver and passenger.

From the initial idea of airbags in 1953 to the modern state of airbags today has been a long journey with many obstacles for chemists and engineers. The development of a functional and safe airbag would not be possible without a firm understanding of chemical reactions and the behavior of gases. Next time you ride in a car think about how the airbag is a prime example of science and technology ultimately benefiting mankind.

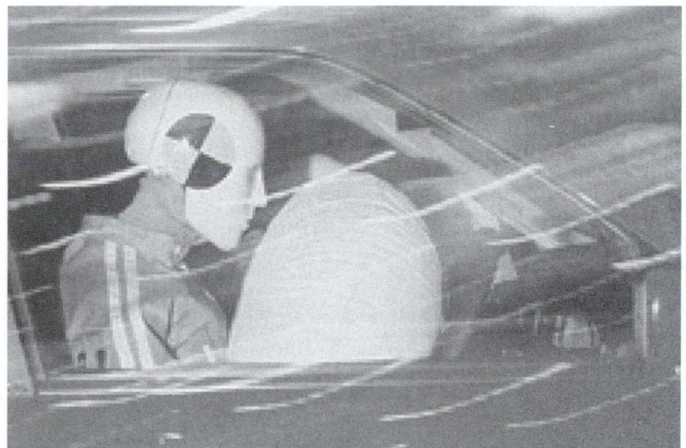

Figure 1. Deployment of an automobile airbag on a test dummy.

BACKGROUND

Airbags

The three chemicals stored in the gas generator of an airbag are sodium azide (NaN_3), potassium nitrate (KNO_3), and silicon dioxide (SiO_2). After a frontal impact of sufficient force, sensing devices located in the front of the car register the impact. The sensing devices send a signal to the generator, where a spark ignites the pellet of sodium azide causing a thermal decomposition (Equation 1). A pre-calculated volume of nitrogen gas is produced and fills the airbag almost completely to capacity (a little more N_2 gas is liberated in the second reaction). The other byproduct of the reaction is hazardous sodium metal, which poses a serious problem due to its reactivity with H_2O.

First Reaction: Triggered by airbag sensor.

$2NaN_3$ (s)	$\rightarrow$	2Na (s)	+	$3N_2$ (g)	Equation 1
sodium azide		sodium metal		nitrogen gas	

The scientists who developed the airbag had to devise a solution to this problem. They found that if potassium nitrate is added to the reaction the sodium can be transformed into a less harmful oxide. Combining the sodium metal with the potassium nitrate makes sense because one of the byproducts is more nitrogen gas (Equation 2).

Second Reaction: Changes the sodium metal into oxides.

Na (s)	+	KNO_3 (s)	$\rightarrow$	K_2O (s)	+	Na_2O (s)	+	N_2 (g)	Equation 2
sodium metal		potassium nitrate		potassium oxide		sodium oxide		nitrogen gas	

The final reaction deals with the remaining oxides. Realizing that the second reaction generates two oxides that could be used to produce a type of glass, the designers introduced silicon dioxide into the composition of airbags. This third reaction happens immediately after the second reaction and results in the formation of alkaline silicate glass (Equation 3). This step was instrumental in the marketing of airbags because there are no toxic materials remaining after deployment of the airbag. The alkaline silicate is very stable and will not ignite.

Third Reaction: Changes the oxides into harmless glass.

K_2O(s)	+	Na_2O(s)	+	SiO_2	$\rightarrow$	alkaline silicate (glass)	Equation 3
potassium oxide		sodium oxide		silicon dioxide			

Gas Stoichiometry

How does our discussion of air bags relate to gas stoichiometry? By knowing the airbag's desired volume, scientists can utilize gas stoichiometry to calculate the quantity of sodium azide required to fill the airbag to near capacity.

As with any problem involving gases, the ideal gas law ($PV=nRT$) will be important. In most gas stoichiometry problems, you are given information (mass, volume or moles) for one substance (A) and must make a conclusion about how much (mass, volume or moles) of another substance (B) can be produced. Take a moment to examine the following flow chart (Figure 2), which summarizes the typical steps in a gas stoichiometry problem. Each double-headed arrow represents one step in solving the problem. The actual number of steps in your calculation will depend on the initial known values and the desired unknown quantity.

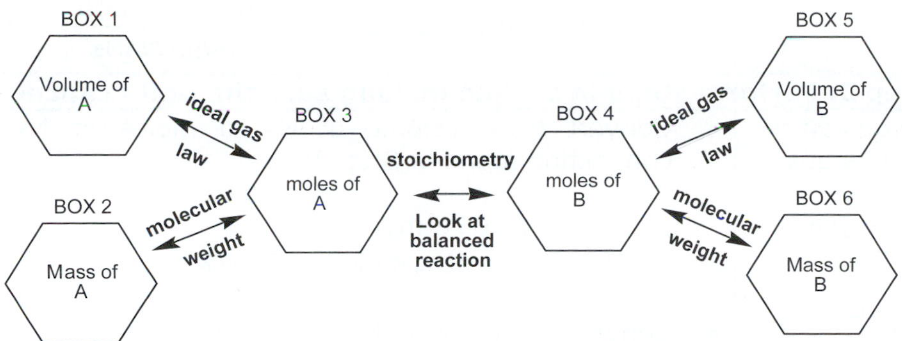

Figure 2. Flowchart for performing gas stoichiometry problems.

Sample problem: How many grams of NaN_3 must be decomposed to fill a 42.0 L airbag with nitrogen gas at a pressure of 1.05 atm and a temperature of 29.5°C?

Sample solution: Table 1 shows a method for solving gas stoichiometry problems. The method is broken down into 4 steps. The column on the left describes what is involved in each step and the column on the right shows the actual work for the sample problem. With a little practice these steps can be generalized and used to solve most dimensional analysis problems.

Table 1: Solution to Sample Problem

Step 1: Organize the problem	
Identify what is **known**.	**Known:** $2NaN_3(s) \rightarrow 2Na(s) + 3N_2(g)$ $V(N_2)=42.0$ L $T(N_2)=29.5°C=302.5$ K $P(N_2)=1.05$ atm R=0.08206 L•atm/K•mol
Identify **unknowns**.	**Unknown:** Moles of NaN_3? Mass of NaN_3?
Make a **prediction**.	**Prediction:** From the balanced chemical equation, the number of moles nitrogen gas produced is greater than the number of moles of sodium azide that decompose.
Step 2: Outline your problem using the flowchart in Figure 2	
Determine starting and ending points on the flow chart. Use these to plan an **outline** for your calculation.	**Outline:** We are given the volume of N_2 needed (corresponds to "volume of A" on the flow chart). This will be converted to moles of N_2 ("moles of A") using the ideal gas law. The moles of N_2 are then converted to moles of NaN_3 ("moles of B") using the stoichiometry from the balanced reaction. The number of grams of NaN_3 ("mass of B") is then determined.
Identify the appropriate **connections** to carry out your outline.	**Connections:** • To convert from volume of N_2 to moles of N_2, ("volume of A" to "moles of A") use: $n_{N2} = \frac{PV}{RT}$ • To convert from moles of N_2 to moles of NaN_3, ("moles of A" to "moles of B") use: $\frac{3 \text{ moles of } N_2}{2 \text{ moles of } NaN_3}$ • To convert from moles of NaN_3 to grams of NaN_3, ("moles

	of B" to "mass of B") use: $\frac{65.02 \text{ g } NaN_3}{\text{mole } NaN_3}$
Step 3: Perform steps in outline making sure the units cancel.	
At this point you can either do all the steps as one, continuous dimensional analysis problem or perform each of the steps individually as shown in the example.	• Convert from volume of N_2 to moles of N_2 using the connection found in Step 2: $\frac{(1.05 \text{ atm})(42.0 \text{ L})}{(0.08206 \text{ L atm/K mol})(302.5 \text{ K})} = 1.78 \text{ moles } N_2$ • Convert from moles of N_2 to moles of NaN_3: $1.78 \text{ moles } N_2 \bullet \frac{2 \text{ moles of } NaN_3}{3 \text{ moles of } N_2} = 1.18 \text{ moles } NaN_3$ • Convert from moles of NaN_3 to grams of NaN_3: $1.18 \text{ moles of } NaN_3 \bullet \frac{65.02 \text{ grams}}{1 \text{ mole of } NaN_3} = \mathbf{77.0 \text{ g } NaN_3}$
Step 4: Check your answer	
Correct units and significant figures?	Yes, the answer has three significant figures.
Is the question answered?	Yes, the question asked to find the mass of NaN_3.
Does the result make sense?	The answer agrees with the initial prediction. We expected the moles of moles N_2 to be greater than the moles of NaN_3.

OVERVIEW

This experiment will be unlike previous procedures, because you will design your own protocol. Your assigned task is to simulate the automobile airbag by filling a plastic bag with gas. However, because sodium azide is extremely explosive and toxic, this experiment will not involve the actual reagents used in airbags. Instead the reaction that will be used to demonstrate gas stoichiometry is the reaction between sodium bicarbonate and acetic acid. This experiment will require detailed note taking, because you must formally write-up your procedure and results as part of your post-lab exercises.

PROCEDURE

Chemicals Used	Materials Used
Sodium bicarbonate 1 M Acetic acid	Small plastic zip lock bag Ruler 50-mL Graduated cylinder Analytical balance and weighing paper Scoopula Compressed air

Your first job as a researcher for an automobile company is trying to devise a new air bag that utilizes less expensive chemicals. You have decided to use the gas generated from the reaction of sodium bicarbonate with acetic acid to fill the plastic bag. As part of your work, you will need to calculate how much of each reagent is required to fill the bag with gas. Test your calculation by attempting to fill the baggie with gas. If the baggie does not fill to capacity (or is over filled), re-examine your calculations.

Your supervisor must know exactly what you did, how you did it, and how much of each reagent was used. Make sure that you take careful notes while performing this experiment.

Caution: When determining the volume of the plastic bag, remember that nothing in lab should be placed in or near your mouth. Since you can't blow up the bag using your lips or a straw, you must find another way to determine the volume of the bag.

Gas Stoichiometry: The Automobile Airbag

Name:	Lab Instructor:
Date:	Lab Section:

PRE-LABORATORY EXERCISES

1. The thermal decomposition of calcium carbonate produces two by-products, calcium oxide (also known as quicklime) and carbon dioxide. Calculate the volume of carbon dioxide produced at STP from the decomposition of 231 grams of calcium carbonate.

2. On December 1, 1783, Jacque Charles became the first human to pilot a non-tethered hydrogen balloon. He flew the balloon for fifteen miles and stayed in the air for a total flight time of 45 minutes. The balloon was made of silk coated with a thin layer of natural rubber and had a diameter of 27 feet. He generated the hydrogen gas needed to lift the balloon by mixing a large amount of iron with aqueous sulfuric acid. If he used 1×10^3 lbs. of iron and excess sulfuric acid, how many liters of hydrogen gas were made?

$$Fe\ (s) + H_2SO_4\ (aq) \rightarrow FeSO_4\ (aq) + H_2\ (g)$$

OVER →

PRE-LABORATORY EXERCISES continued...

3. Another process that is used to inflate things such as weather balloons, life rafts, etc. is the reaction between calcium hydride and water to give calcium hydroxide and hydrogen gas. How many grams of calcium hydride are needed to fill a life raft having a volume of 10.2 L and a pressure of 735 torr at 22°C?

4. Write the balanced chemical reaction that will be used to fill the air bag you make in this experiment. Hint: One of the products of the reaction between sodium bicarbonate (also called sodium hydrogen carbonate) and acetic acid is a gas.

Gas Stoichiometry: The Automobile Airbag

Name:	Lab Instructor:
Date:	Lab Section:

RESULTS and POST-LABORATORY QUESTIONS

For this experiment instead of having post-lab questions, you are required to write a laboratory report to prepare you for the type of work required in subsequent science courses. Unless instructed otherwise, follow the format given below for preparing your typed laboratory report.

I. Introduction
   a) Describe the goal of the experiment.
   b) Introduction to airbags (include a brief history of airbags, the purpose of airbags and how they function). For this section you will need to summarize an article on airbags. Record the reference for the article or the URL of the website where you found the information along with the date when you accessed the site.

II. Experimental
   a) Write the procedural protocol that you devised. Make sure to include a list of materials, the quantities of reagents and a detailed stepwise procedure. Include a description of how to determine the volume of the bag.

III. Results and Discussion
   a) Following the model in Table 1 in the background, show the gas stoichiometry calculation for your airbag.
   b) Describe the pros and cons of the particular chemical reaction used to inflate the airbag.
   c) Comment on the number of trials preformed, and how the experiment could be improved along with any experimental errors you made.

Introduction to Thermochemistry: Using a Calorimeter

D. Van Dinh

OBJECTIVES

- Investigate heat flow between a system and its surroundings.
- Determine the heat of reaction using a simple coffee-cup calorimeter.
- Correlate the magnitude of the ΔH_{rxn} for an acid-base neutralization with the strength, identity, and concentration of the acid.

INTRODUCTION

Have you seen advertisements for the "revolutionary defrosting tray" called Miracle Thaw™? It thaws food in a matter of minutes without electricity, chemicals or batteries (Figure 1). Miracle Thaw™ is a Teflon-coated aluminum tray that achieves accelerated thawing of frozen food by absorbing the heat energy in the surrounding air and releasing it directly into the frozen food. This product illustrates the concept of energy flow by showing how a metal can transfer heat to another object. In doing so, the Miracle Thaw™ causes a physical change in the frozen food. All changes in matter, whether chemical or physical, are accompanied by changes in energy, either in the form of heat or work. In this lab, you will learn how to measure the quantity of heat released or absorbed in a reaction. For example, how do we determine the amount of heat the ice cube in Figure 1 is absorbing from the surroundings as it melts?

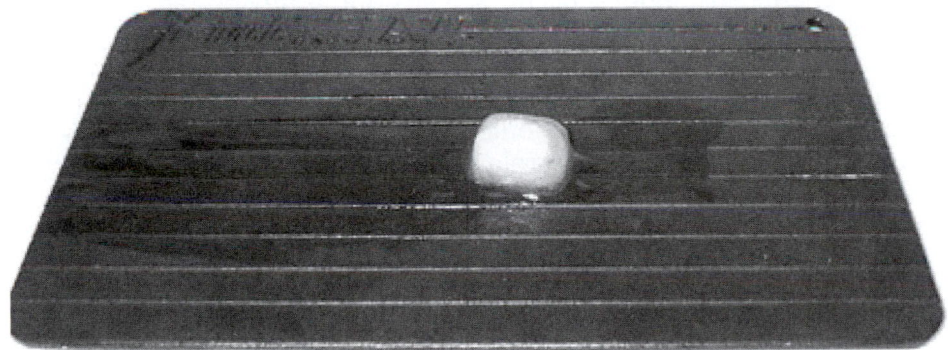

Figure 1. An ice cube begins melting immediately upon contact with Miracle Thaw™.

BACKGROUND

Thermochemistry

Energy is the capacity to do work or transfer heat. Energy may be classified as either potential or kinetic and though energy can be converted from one form to another, the total energy content of the universe is constant.

All chemical and physical changes, whether melting ice or burning fuel in a car, are accompanied by changes in the internal energy, ΔE, of the system (Equation 1).

$$\Delta E = E_{final} - E_{initial} \qquad \text{Equation 1}$$

A change in the energy of the system is always accompanied by an opposite change in the energy of the surroundings. For example, in the case of the combustion of gasoline, the internal energy of the final products (CO_2 and H_2O) is lower than the internal energy of the initial gasoline resulting in a negative value of ΔE. Unfortunately, it is not possible to easily determine the internal energy of any systems of practical interest. Luckily, we have another method to do this, since the energy transfer outward from the system or inward from the surroundings appears in only two forms, heat and work. Since energy can be transferred as heat, q, and/or work, w, the total change in a system's internal energy is $\Delta E = q + w$. Figure 2 illustrates the agreed upon convention for the signs of q and w. When heat is added to the system (endothermic) or work is done on the system, the signs of q and w are positive. When heat is removed from the system (exothermic) or work is done by the system, the signs of q and w are negative.

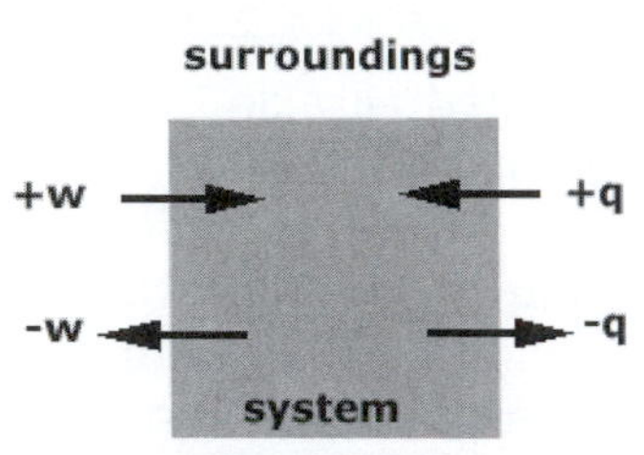

Figure 2. Illustration of the sign convention of work (w) and heat (q)

Calorimetry

Most physical and chemical changes occur at constant atmospheric pressure, for example a reaction in an open flask or the freezing of a lake. The heat gained or lost at constant pressure is termed the heat of reaction or the enthalpy change, ΔH. Because it is this heat that we will measure in this experiment, we should ask, "How do scientists measure the *amount* of heat absorbed or released by a system during a physical or chemical process?"

The study of calorimetry involves the measurement of heat changes during physical and chemical processes. A calorimeter is an insulated vessel used to measure the heat exchange between the system being observed and its surroundings during a process. A simple calorimeter can be constructed from two nested Styrofoam cups, a lid, and a thermometer (Figure 3).

Figure 3. Typical coffee-cup calorimeter.

To determine the heat flow, q_{rxn}, for the reactions in today's experiments, the calorimeter will contain an acid-base neutralization reaction (the system) in contact with 60 mL of water (the surroundings). Heat released by the system will warm the 60 mL of water within the calorimeter and the temperature change (ΔT) of the water will be measured. But how does knowing ΔT for the water allow us to determine the amount of heat, q_{H2O}, absorbed or released by the water? The energy associated with the temperature change in a substance is often expressed in terms of specific heat capacity. You know from everyday experience that the more heat an object absorbs, the hotter it gets (think of your car sitting in the sun). Every object has its own particular capacity for absorbing heat. It may seem strange that adding equivalent amounts of heat to two different objects (Ex: adding 1 Joule of heat to 1 g H_2O vs. 1 g Fe) would result in different ΔT, but that is exactly what we find experimentally. For example, the specific heat of water is 1.00 cal/g°C (4.18 J/g•K) and the specific heat of Fe is 0.108 cal/g°C (0.450 J/g•K). The heat gained or lost by the water is related to its specific heat according to Equation 2.

$$q_{H_2O} = (\text{mass of water})(\text{specific heat of water})(\Delta T) \qquad \text{Equation 2}$$

We can now put together all the pieces of the puzzle. Remember, our goal is to calculate the heat of the reaction, q_{rxn}. As shown in Equation 3, since all of the heat released by the reaction must be gained by the water or the calorimeter, we can set q_{rxn}, equal to the negative of the heat gained by the water, q_{H_2O} (from Equation 2), plus that gained by the calorimeter, q_{cal}.

$$q_{rxn} = -\ (q_{H_2O} + q_{cal}) \qquad \text{Equation 3}$$

In this experiment, we will assume that q_{cal} will be zero so we have:

$$q_{rxn} = -\ q_{H_2O} \qquad \text{Equation 4}$$

By calculating q_{H_2O} (according to Equation 2), we are able to determine q_{rxn}. In this laboratory experiment, the temperature change, ΔT, undergone by the 60 mL of water will be used to calculate the q_{H_2O}. To determine the mass of the water, assume that the density of each solution is equivalent to the density of pure water. Also, assume that the specific heat of the solution is the same as the specific heat of water. It should be noted that the final value for q_{rxn} is the same thing as the enthalpy change, ΔH, that we discussed earlier.

OVERVIEW

To help you understand calorimetry, in Part A you will measure the amount of heat needed to sublime a specific quantity of dry ice. In Part B, you will work in pairs to study the heat of reaction associated with various acid-base neutralization reactions. Each pair will be assigned a different acid. After completing the experiment, each pair will share their results with the class.

PROCEDURE

Part A: Measuring the Amount of Heat Needed to Sublime Dry Ice

Chemicals Used	Materials Used
Dry Ice (20-25 g)	Calorimeter 50-mL Graduated Cylinder Thermometer Hot Plate 100 and 300-mL Beakers

Assemble the calorimeter according to your instructor's directions. Heat 40.0 mL of de-ionized H_2O in the 100-mL beaker. When the temperature of the water is between 70-80°C, take the beaker off the hot plate. Pour the warm H_2O into the calorimeter and record the temperature. Place about 15 g of dry ice in a tared 30-mL beaker. Record the exact mass of ice. Pour the dry ice into the calorimeter and stir. Record the temperature of the "warm" water when all of the dry ice has sublimed. Calculate the amount of heat needed to sublime one mole of dry ice and write your result on the board with the class data.

Part B. Heat of Reaction for an Acid- Base Reaction

Chemicals Used	Materials Used
2.2 M NaOH (30.0 mL) 2.0 M HCl or 2.0 M HNO_3 (30.0 mL)	Calorimeter 50-mL Graduated Cylinder Thermometer Various beakers

CAUTION: NaOH, HCl, and HNO_3 are corrosive. If any chemicals are spilled on the skin, wash them off immediately with water and inform your instructor.

Transfer 30.0 mL of 2.2 M NaOH solution to a clean, dry calorimeter. Allow the solution to reach room temperature (about 3 minutes) and record the temperature with a clean, dry thermometer. Add 30.0 mL of either 2.0 M HCl or 2.0 M HNO_3 solution (depending on which acid your group has chosen) to a clean, dry graduated cylinder. Allow the solution to reach room temperature (about 3 minutes), and record the temperature of the solution with a clean, dry thermometer. Pour the acid solution into the calorimeter, stir well, and record the temperature at 15 second intervals for 2 minutes. Flush the resulting salt solution down the sink with running water. Calculate your value for the heat of reaction, ΔH_{rxn}, and write your result on the board with the class data.

Introduction to Thermochemistry: Using a Calorimeter

Name:	Lab Instructor:
Date:	Lab Section:

PRE-LABORATORY EXERCISES

1. Define the underlined terms in the BACKGROUND section.

2. Determine whether each of the following processes is exothermic or endothermic.
   a) The freezing of water to make ice-cubes.

   b) The digestion of the fatty acids contained in the oil of your salad dressing

   c) The evaporation of sweat during exercise.

3. You have just moved into a new apartment and you need some pots and pans to cook with. Using your knowledge of specific heat capacity, which kind of pots and pans would you purchase: ones made of glass, copper, or iron? Explain. (You may have to use the Internet or another textbook to find the specific heat capacities of these three substances. If so, record the URL of the web site or the name of the book where you found the information.)

OVER →

PRE-LABORATORY EXERCISES continued...

4. How many kilocalories are required to change the temperature of 80.0 g of water from 23.3°C to 38.8°C?

5. The calorimeter used in this experiment consists of Styrofoam cups. What is the advantage of using Styrofoam cups instead of glass or paper cups?

6. The experimental procedure in Part B has you wash your thermometer and dry it after you measure the temperature of the NaOH solution and before you measure the temperature of the HCl solution. Why?

7. Write the balanced molecular and net ionic equations for both of the acid-base neutralization reactions described in Part B.

Introduction to Thermochemistry: Using a Calorimeter

Name:	**Lab Instructor:**
Date:	**Lab Section:**

RESULTS and POST-LABORATORY QUESTIONS

Part A. Measuring the Amount of Heat Needed to Sublime Dry Ice

Volume, warm water (mL)	
ΔT, warm water (°C)	
q for warm water (cal)	
q for dry ice (cal)	
Mass, dry ice (g)	
Heat needed to sublime 1 mol dry ice (kcal/mol)	
Heat needed to sublime dry ice, class average	

1. Identify the system and surroundings in Part A.

2. Are we directly measuring a temperature change in the system or the surroundings?

3. Based on the temperature change measured, did the surroundings gain or lose heat? Explain.

4. Based on your answer to #3 above did the system gain or lose heat? Explain.

5. What is the sign of q for the sublimation of dry ice? Explain.

6. Is the sublimation of dry ice exothermic or endothermic? Explain.

7. Show your work for calculating q_{H_2O} for the warm water.

8. The heat needed to sublime 1 mol of dry ice (our q_{rxn}) is also called $\Delta H_{sublimation}$. Show your work for calculating your $\Delta H_{sublimation}$.

9. Find the value for $\Delta H_{sublimation}$ for dry ice. Calculate the percent error based on your class average.

RESULTS and POST-LABORATORY QUESTIONS continued...

Part B. Heat of Reaction for a Strong Acid-Strong Base Reaction

	HCl or HNO_3 (circle the acid you used)	HCl or HNO_3 (circle the other acid)
Total volume of water in calorimeter (mL)		
Avg. initial temp. of acid and base (°C)		
Highest temp. of solution after reaction (°C)		
ΔT, water (°C)		
Heat gained or lost by water, q_{H2O} (cal)		
Heat of reaction, q_{rxn} (cal)		
Concentration of acid		
Heat of reaction, q_{rxn} (kcal/mol acid)		
Heat of reaction, class average		

1. Identify the system and surroundings in Part B.

2. What did the temperature change you measured tell you about the ΔT for the system? Explain.

3. Are acid-base neutralizations exothermic or endothermic? Explain.

4. Show your work for calculating the heat of solution, q_{H2O}.

5. Show your work for calculating the molar heat of reaction, q_{rxn}, in kcal/mol.

6. Did the identity of the strong acid (HCl versus HNO_3) affect the molar heat of reaction, q_{rxn}? Justify your answer in terms of the net ionic reactions you wrote in PRE-LABORATORY EXERCISE #7.

7. In our experiment, we assumed the heat capacity of the calorimeter to be zero. Based on this assumption, would you expect your calculated results to be less than or greater than the actual ΔH_{neut} if we had calculated the heat capacity of the calorimeter? Explain.

Calorimetry: Nutrition in a Nutshell

Kristen Spotz

OBJECTIVES

- Explore a practical application of calorimetry.
- Determine the heat of combustion of a sample of peanuts.

INTRODUCTION

On a recent trip to the San Diego Zoo, one of the authors of this laboratory manual saw a small display of interest near the macaque (a type of small monkey) cage. The display contained an enclosed bomb calorimeter (Figure 1) along with the explanation that the zookeepers had performed studies to know how many Calories each macaque required daily for a healthy diet. The display went on to explain that the zookeepers take various samples of Macaque food (such as small caterpillars) and using a bomb calorimeter, like the one displayed, determine the number of Calories per sample. From there it is an easy step to figure out how many caterpillars to feed each monkey each day. A similar process is used in the labeling of the Calorie content on the food you eat. In this laboratory experiment you will calculate the Calorie content in a sample of peanuts using a simplified (non-bomb) version of the calorimeter used by food scientists.

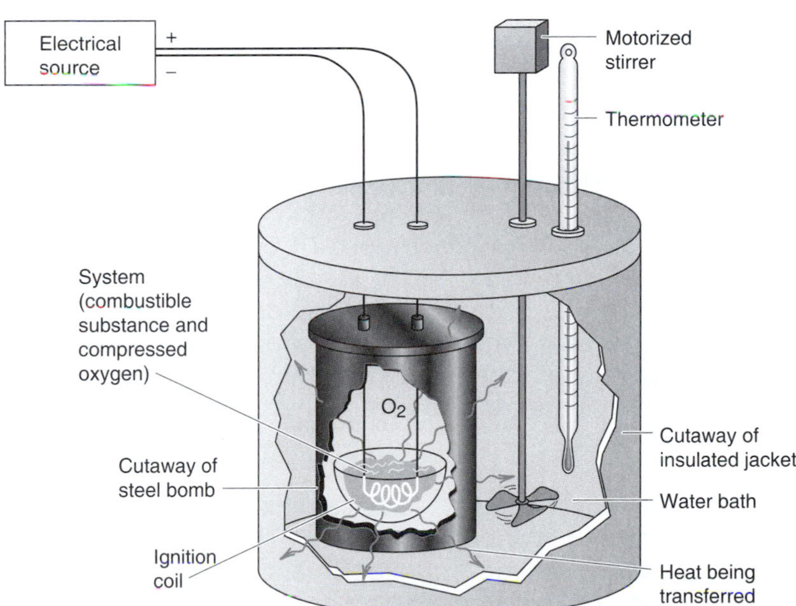

Figure 1. A bomb calorimeter like the one used to ensure a healthy diet for the animals at the zoo.

BACKGROUND

Nutrition in a Nutshell

During photosynthesis, the light energy from the sun is stored in green plants. As we eat the plants, or we eat animals that eat the plants, our body converts that energy into a form that we can then use. The nutrients from the plants are broken down during a process similar to combustion. The energy released during this process is utilized by our bodies to perform work. The specific amount of energy released depends on the chemical composition of the different foods.

Peanuts, for example contain all three major macronutrients (proteins, carbohydrates, and lipids) in plentiful amounts and are a great source of energy. On average, the **gross energy content** for lipids (fats and oils), carbohydrates, and proteins are 9.40 Cal/gram, 4.15 Cal/gram and 5.65 Cal/gram respectively. A 1.000-gram portion of peanuts contains roughly 0.484 grams of fat, 0.259 grams of protein, and 0.164 grams of carbohydrates (the remaining mass does not contribute to the energy content). The total gross energy content of the 1.000-gram portion of peanuts can be calculated as in Equation 1:

$$\text{Gross energy} = (0.484 \text{ g} \times 9.40 \text{ Cal/g}) + (0.259 \text{ g} \times 5.65 \text{ Cal/g}) + (0.164 \text{ g} \times 4.15 \text{ Cal/g})$$

$$= 6.69 \text{ Cal} \qquad \text{Equation 1}$$

Note that while approximately 50% of a peanut's mass consists of lipids, this fat is responsible for nearly 70% of the peanut's energy content. The carbohydrates and proteins contribute the remaining 30%. More calories are derived from fat than from carbohydrates, due to the differences in chemical composition of the two types of materials. For example, lauric acid, $C_{12}H_{24}O_2$ (a fatty acid), and sucrose $C_{12}H_{22}O_{11}$ (a carbohydrate), have the same number of carbon atoms and nearly the same number of hydrogen atoms. During the metabolism of these compounds, the carbon and hydrogen atoms combine with oxygen to produce the end products of metabolism, CO_2 and H_2O (Equations 2 and 3). Since fats have fewer C-O bonds to start with, they release more energy than carbohydrates during the metabolism process.

Lipid: $C_{12}H_{24}O_2 + 17\ O_2 \rightarrow 12\ CO_2 + 12\ H_2O$ Equation 2

Carbohydrate: $C_{12}H_{22}O_{11} + 12\ O_2 \rightarrow 12\ CO_2 + 11\ H_2O$ Equation 3

The chemical change associated with burning peanut oil ($C_{57}H_{104}O_6$) involves breaking the high-energy bonds between the carbon, hydrogen, and oxygen within the fatty acids. As these bonds are broken, energy is given off into the surroundings. This overall process causes a decrease in the energy content of the resulting products.

Calorimetry

As explained earlier, the bonds of the fatty acids, amino acids or carbohydrates in food are broken to release heat (energy) during digestion. This process is very similar to the chemical process of combustion. In today's experiment, we will measure the gross energy content of a sample of food, by burning the food and collecting the released heat. The collection and measurement of heat involves the use of a calorimeter. The calorimeter utilized in this experiment specifically measures the change in the amount of heat released or absorbed by a system at constant pressure, a quantity known as enthalpy. By using water as a collecting medium for the released heat, we can measure the heat given off by the burning food. The value of the specific heat indicates how many joules of heat are required to raise the temperature of 1 g of a substance by 1 K. Since the specific heat of water is known (4.184 J/gK), along with the mass of water in the calorimeter, we can calculate the amount of heat absorbed by the water if we measure the temperature change of the water (Equation 4).

$$q_{H2O} = (\text{mass of water})(\text{specific heat of water})(\Delta T) = -\, q_{combustion} \qquad \text{Equation 4}$$

The heat given off by the system during combustion ($q_{combustion}$) must be equal to the negative of the heat taken in by the surroundings (q_{H2O}). The mass of water is given in grams. The change in temperature is defined as the final temperature minus the initial temperature of the water.

OVERVIEW

After constructing a simple soda can calorimeter, you will calibrate the efficiency of the calorimeter by burning a votive candle, which has a known enthalpy of combustion. Next, you will use your calorimeter to measure the amount of energy released during the combustion of a peanut.

PROCEDURE

Chemicals and Materials Used	
Peanuts (or other nuts)	Ring stand and 2 iron rings
Votive candle	Bunsen burner
2 aluminum soda cans (empty)	Flame igniter or matches
100-mL graduated cylinder	Tongs
Thermometer	Scissors
De-ionized water	Analytical balance

Part A. Building and Calibrating a Calorimeter

1. Using scissors, completely cut one of the aluminum cans 1 inch from the bottom. This will serve as the fuel container. Without cutting through the bottom of the fuel container, make a small indention in the center of the fuel container to serve as a well.

2. Cut a second aluminum can $1^1/_2$ inches from the top. Create two flaps to allow the container to rest on the iron ring. This will serve as the water container.

3. Completely dry the inside of the fuel and water containers. Obtain a ring stand and attach two iron rings. Position the water container in the top ring with the thermometer inside (Figure 2).

4. Use a graduated cylinder to transfer 100.0 mL of de-ionized water to the water container. Stir the water using the thermometer and measure the temperature (± 0.2°C).

Figure 2. Set up of a soda can calorimeter

5. Weigh a votive candle to the nearest milligram. Light the candle and quickly place it in the fuel container in the bottom iron ring. Adjust the height of the fuel container, so it is as close as possible to the water container without smothering the flame. Your water container should touch the tip of the flame.

6. Continue to stir the water with the thermometer. Once the temperature has increased by 10 - 12°C above your starting temperature, extinguish the flame. Record the final temperature (± 0.2°C).

7. Allow the calorimeter to cool for 2 – 3 minutes and reweigh the candle. Empty the water and thoroughly dry the entire apparatus.

8. Repeat steps 4 - 7 above. If the percent difference in efficiency (Equation 5) between your first and second trials is greater than 1%, repeat a third time. See RESULTS and POST-LABORATORY QUESTIONS for Part A on how to calculate the efficiency.

$$\text{Percent difference} = \frac{(\text{Efficiency, trial 1}) - (\text{Efficiency, trial 2})}{(\text{Efficiency. trial 1})} \times 100 \qquad \text{Equation 5}$$

Part B. Measuring the Heat of Combustion of Nuts

9. Obtain two peanut halves and place them into the fuel container. Record the weight of the fuel container and the peanuts to the nearest milligram.

10. Using the graduated cylinder, add 100.0 mL of de-ionized to the water container of the calorimeter. Stir the water with the thermometer and measure the initial temperature (± 0.2°C).

11. Light the bunsen burner and, using the tongs, place the peanut halves in the flame. Once the peanut ignites, keep it in the flame for an additional 4 – 5 seconds to make sure it is fully burning. Transfer the peanut into the fuel container and raise the level of the iron ring until the flame touches the water container. This step may take some patience in order to keep the peanut lit. You should also make sure you are not in a drafty area.

12. Using the thermometer, stir the water. Immediately after the peanut has stopped burning, measure the temperature of the water (± 0.2°C). After the fuel container has fully cooled (2 - 3 minutes), record the mass of the peanut residue and the fuel container.

13. Using the calculated efficiency of your calorimeter and data obtained in this part of the experiment, calculate the enthalpy of combustion for the burned mass of the peanut.

14. Allow the calorimeter to cool completely before emptying the water and disposing of the peanut.

15. Repeat steps 9 – 14 for a second trial.

16. Optional: Repeat this procedure with a different nut (almonds, walnuts, cashews, etc...).

Calorimetry: Nutrition in a Nutshell

Name:	Lab Instructor:
Date:	Lab Section:

PRE-LABORATORY EXERCISES

1. Food labels normally report information in terms of Calories (Cal, with a capital "C") where 1 Cal = 1000 cal = 1 kcal. However, the S.I. unit of expressing energy is the joules (J). If one pound of pure fat yields about 4.2 $x10^3$ kcal when burned, convert this value to Calories, calories and joules.

2. Assuming no calorie intake from water or minerals, if a 100.-gram steak contains 49% water, 15% protein, 0% carbohydrate, and 36% fat what is the estimated gross energy content of the steak (in Calories)?

   For every 3500 Cal consumed but not burned up through activity, roughly 1 pound of body fat is stored. If no energy is burned up during the eating of the steak, how many pounds of body fat are stored?

3. The burning of 15.0 grams of vegetable oil causes the temperature of 250.0 mL of water in the calorimeter to rise from 22.5ºC to 33.0ºC. Determine the amount of heat given off during this combustion process (in Cal/g). Assume the calorimeter is 100% efficient.

OVER →

PRE-LABORATORY EXERCISES continued...

4. The human body is not a 100% efficient system and we are unable to harness all the energy released from the metabolism of foods. The amount of energy actually harnessed by the body represents the **net energy value** (also called the **physiologic energy value** or the **food energy value)** and is the value that appears on food labels. The average net energy value for lipids is 9.00 Cal/gram and for carbohydrates and proteins the average net energy value is 4.00 Cal/gram. Calculate the net energy value of the 1.0-g portion of peanuts described in the BACKGROUND section.

   Compared to the gross energy value found in Equation 1, what is the % efficiency with which human body harnesses the energy of the peanuts?

5. Suppose you had two peanuts of equal mass and burned each in a separate calorimeter. The only difference between the calorimeters is the volume of water as indicated in the picture below.

A: 25 mL of water

B: 150 mL of water

   Which calorimeter (A or B) will exhibit the greatest change in temperature? Explain.

   Will changing the volume of water affect the results of the experiment in terms of the Cal/g of peanut burned? Explain.

Calorimetry: Nutrition in a Nutshell

Name:	Lab Instructor:
Date:	Lab Section:

RESULTS and POST-LABORATORY QUESTIONS

Part A. Building and Calibrating a Calorimeter

	Trial 1	Trial 2	Trial 3 (optional)
Mass of water, g			
Δ Temp. for water, K			
Mass of candle burned, g			
Heat absorbed by the water (q_{H2O}), J #			
Heat absorbed by the water (q_{H2O}), kcal			
Theoretical heat, kcal ##			
Efficiency of calorimeter ###			
Average efficiency (use two trials within 1%)			

\# q_{H2O} = (mass of water)(specific heat of water)(ΔT)

\## Theoretical heat produced by burning candle = (mass candle burned) x (10.0 kcal/g)

\### Efficiency of calorimeter = $\dfrac{\text{Heat absorbed}}{\text{Theoretical heat yield}}$

You should end up with a positive value for q_{H2O}. What does that tell you about the sign of q for the combustion of the candle? Is the combustion process exothermic or endothermic? Explain.

RESULTS and POST-LABORATORY QUESTIONS continued...

Part B. Measuring the Heat of Combustion of Nuts

	Trial 1	Trial 2
Mass of water, g		
Δ Temp. for water, K		
Mass of peanut burned, g		
Heat absorbed by the water (q_{H2O}), J		
Heat absorbed by the water (q_{H2O}), Cal		
Enthalpy of combustion for nut, Cal #		
Enthalpy of combustion per gram of nut, Cal/g		
Avg. Enthalpy of combustion, Cal/g		

\# Calculate the corrected enthalpy of peanut combustion by using the average calorimeter efficiency from Part A.

$$\Delta H = \frac{\text{heat absorbed by water (Cal)}}{\text{average efficiency of calorimeter}}$$

Explain why you stirred the water in the calorimeter throughout the heating process until the final temperature reading.

Compare the gross energy yield for a 1.0 g peanut found in Equation 1 in the BACKGROUND to your experimental value for the enthalpy of combustion per gram of burned peanut. Relate any difference in these values to possible experimental errors. What could you do to help reduce any experimental errors?

Hess's Law: A Study of the Combustion of Magnesium

Holly Morrison

OBJECTIVE

- Perform calculations involving Hess's law.
- Use a calorimeter to determine the heat of reaction for a chemical process.

INTRODUCTION

To appreciate Hess's law, it is important to have a conceptual understanding of what is meant by a state function. Consider the following scenario: Both you and your best friend from high school will be driving your cars this summer from your homes in New York to attend college in California. You plan on taking a fairly straight path to California while your friend plans on making a detour to visit his grandparents in Miami. The trip from New York to California is an example of a state function. Even though you will have taken two separate paths (and he will have driven more miles than you), you and your friend will travel the same absolute distance (the displacement) because you both started and ended in the same place. In this sense your road trip is analogous to a chemical reaction where the reactants are the starting place and the products are where you end up. For chemical reactions, the enthalpy change is a state function that is independent of the actual path taken to get from reactants to products.

BACKGROUND

Hess's Law

In experiments conducted with a calorimeter, the enthalpy change for a reaction can be measured. Chemists are often presented with a challenge, however, because not all reactions take place in a single step. Many chemical reactions are comprised of several interwoven steps, and it is impossible to examine each individually. In 1840, Swiss chemist Germain Henri Hess published his law of heat summation, which is now called Hess's law. The law states that "the enthalpy change of an overall process is the sum of the enthalpy changes of its individual steps." In other words, because ΔH is a state function, the enthalpy of a reaction can be determined by considering the reaction as a sum of individual reaction steps that are chosen because they have known values of ΔH. In this approach, related reactions can be studied and measured in order to find the enthalpies of those we can't measure directly. This discovery paved the way for the study of thermodynamics in the 19th century and allows us to get around the problem of determining the enthalpy change for a reaction that we can't run in a laboratory.

Calculating the ΔH of a target reaction using Hess's law is a four-step process. First, we must have reactions of known enthalpies that can be manipulated to generate the target reaction. Typically these reactions will be supplied for you. Second, using clues in the target reaction (such as the number of moles of each species and whether each species is a reactant or product) we manipulate the given reactions so that they can be summed to give the target reaction. Next, we perform the same manipulations with the corresponding values of ΔH that we were provided with. For example, if the reaction needs to be reversed to generate the goal reaction, then the sign of the ΔH must be changed or if one quantity of moles is multiplied by two, the same must be done for all of the other reactants and products of that specific reaction as well as the ΔH for that reaction. Finally, we add the manipulated reactions to obtain the target reaction and we add the manipulated quantities of ΔH to obtain the unknown value of ΔH for the target reaction.

An Example Using Hess's Law

To appreciate the significance of Hess's law, let's imagine that a scientist studying the ozone hole needs to know the heat of reaction, ΔH, for the reaction of hydrogen gas with ozone to produce water vapor (Equation 1). She checks the research literature and finds that no one has previously determine the ΔH for this process. She also does not have the equipment to perform a calorimetry experiment to measure the ΔH herself. With a little detective work, however, she is able to find two related reactions (Equations 2 and 3) that, when manipulated and added together, result in the goal reaction. According to Hess's law, our scientist knows that similar manipulation of the known ΔH values will result in the desired unknown ΔH value for Equation 1.

$$3\,H_2\,(g) + O_3\,(g) \rightarrow 3\,H_2O\,(g) \qquad \Delta H = ??? \qquad \text{Equation 1}$$

$$2\,H_2\,(g) + O_2\,(g) \rightarrow 2\,H_2O\,(g) \qquad \Delta H = -\,483.6\text{ kJ} \qquad \text{Equation 2}$$
$$3\,O_2\,(g) \rightarrow 2\,O_3\,(g) \qquad \Delta H = +\,284.6\text{ kJ} \qquad \text{Equation 3}$$

Looking at our goal reaction (Equation 1) we need 3 moles of gaseous H_2 on the reactant side. The only place that H_2 gas appears in Equations 2 and 3 is as a reactant in Equation 2. That the H_2 is already a reactant in Equation 2 indicates that we will not have to flip the reaction. However, there are only 2 moles of H_2 in Equation 2 so we will multiply the entire reaction (as well as the corresponding ΔH) by 3/2 to give the needed 3 moles of H_2 gas (Equation 2a).

Manipulation of Equation 3 requires not only dividing the reaction (and ΔH) by 2, but also reversing the reaction (and taking the negative sign of the ΔH) in order to give the required 1 mole of O_3 on the reactant side of Equation 1. The resulting reaction is given in Equation 3a.

$3\,H_2\,(g) + 3/2\,O_2\,(g) \rightarrow 3\,H_2O\,(g)$	ΔH = - 725.4 kJ	Equation 2a
$O_3\,(g) \rightarrow 3/2\,O_2\,(g)$	ΔH = - 142.3 kJ	Equation 3a

At this point our scientist is able to add Equations 2a and 3a together. The 3/2 moles of O_2 cancel to give Equation 1a along with the desired, previously unknown value of ΔH.

$3\,H_2\,(g) + O_3\,(g) \rightarrow 3\,H_2O\,(g)$	ΔH = - 867.7 kJ	Equation 1

The Combustion of Magnesium

Magnesium oxide has several uses in industry. It is produced for use as an animal feed supplement, a pigment extension in paint and varnish, an ingredient in certain types of cements, a semiconductor for electrical wires, and as a component in scientific and decorative glassware. It is generally obtained by processing naturally occurring minerals like magnesium carbonate and brine. One common way to produce magnesium oxide in the laboratory is to burn thin strips of magnesium in the presence of excess oxygen.

In this experiment, our goal will be to determine the heat of combustion by which magnesium and oxygen generate magnesium oxide (Equation 4).

$Mg\,(s) + \frac{1}{2}\,O_2\,(g) \rightarrow MgO\,(s)$	ΔH = ???	Equation 4

As with the scientist in the previous example, imagine that we are unable to find the desired ΔH value in the research literature (in fact you will see in the RESULTS and POST-LABORATORY QUESTION #4 that it is available in your textbook). Although we have the equipment in the laboratory to perform a basic calorimetry experiment, it is difficult to directly obtain the needed information using the available coffee-cup calorimeters. We can, however, follow our scientist's lead by using Hess's law to manipulate and add together the appropriate reactions (Equations 5 - 7). Equations 5 and 6 correspond to the reactions of magnesium and magnesium oxide with hydrochloric acid respectively. Unlike for the combustion of magnesium, the ΔH for these two reactions can easily be determined using a coffee-cup calorimeter. Equation 7 corresponds to the standard enthalpy of formation for water vapor [ΔH_f^o, H_2O (g)] and can be found in your textbook.

$Mg\,(s) + 2\,HCl\,(aq) \rightarrow MgCl_2\,(aq) + H_2\,(g)$	ΔH = experiment	Equation 5
$MgO\,(s) + 2HCl\,(aq) \rightarrow MgCl_2\,(aq) + H_2O\,(g)$	ΔH = experiment	Equation 6
$H_2\,(g) + \frac{1}{2}\,O_2\,(g) \rightarrow H_2O\,(g)$	ΔH = textbook	Equation 7

OVERVIEW

In this experiment, you will determine the ΔH for the reaction of hydrochloric acid with magnesium metal and magnesium oxide powder. The results of these two experiments will be used to calculate the heat of combustion for magnesium using Hess's law.

PROCEDURE

Chemicals Used	Materials Used
1 M HCl Magnesium ribbon Magnesium oxide powder	Coffee-cup calorimeter (2 Styrofoam cups nested in a beaker) Analytical balance Weighing paper Thermometer 25-mL Graduated cylinder Scoopula

1. Cut off approximately 0.1 gram of magnesium ribbon. Record the actual mass of Mg.

2. Weigh a clean, dry coffee-cup calorimeter. Add 25 mLs of 1 M HCl to the calorimeter. Record the mass of the calorimeter and HCl (aq). Record the temperature of the HCl solution.

3. Drop the magnesium ribbon into the HCl. While stirring, record the temperature at 15-second intervals for 2-3 minutes (until the temperature reaches a maximum).

4. Repeat steps 1 – 3.

5. Repeat steps 1 - 4 using approximately 0.25 grams of MgO powder. Again, record the actual mass of the MgO.

6. When you are done cleaning up your experiment and disposing of waste solutions according to your instructors directions, complete the first 6 rows in the results table (RESULTS and POST-LABORATORY QUESTIONS). Write your average values for the heat of reaction for Mg and MgO with HCl on the board with the class data.

Hess's Law: A Study of the Combustion of Magnesium

Name:	Lab Instructor:
Date:	Lab Section:

PRE-LABORATORY EXERCISES

You will need to perform calorimetry calculations during this experiment. Before coming to laboratory, check your textbook for a detailed account on performing these calculations.

1. Define the underlined words in the BACKGROUND section.

2. Which of the following aspects of a drive across country are state functions? Explain.
   a) Days spent

   b) Overall change in altitude

   c) Gallons of gas used

3. Discuss why it would be difficult to measure the heat of reaction for the combustion of magnesium (Equation 4) directly using a coffee-cup calorimeter.

4. Record the value for the standard enthalpy of formation of water vapor (Equation 7) in your textbook. Also record the value in your laboratory notebook so you will have it available while completing the experiment.

OVER →

PRE-LABORATORY EXERCISES continued...

5. Use Hess's law to calculate the heat of reaction for the formation of methanol from methane:

   $2CH_4\ (g) + O_2\ (g) \rightarrow 2CH_3OH\ (l)$ $\Delta H = ???$

   from:

   $CH_4\ (g) + H_2O\ (g) \rightarrow CO\ (g) + 3H_2\ (g)$ $\Delta H = +\ 206.10\ kJ$
   $2H_2\ (g) + CO\ (g) \rightarrow CH_3OH\ (l)$ $\Delta H = -\ 128.33\ kJ$
   $2H_2\ (g) + O_2\ (g) \rightarrow 2H_2O\ (g)$ $\Delta H = -\ 483.64\ kJ$

Hess's Law: A Study of the Combustion of Magnesium

Name:	Lab Instructor:
Date:	Lab Section:

RESULTS and POST-LABORATORY QUESTIONS

	Heat of Reaction: Mg and HCl (Equation 5)		Heat of Reaction: MgO and HCl (Equation 6)	
	Trial 1	Trial 2	Trial 1	Trial 2
Mass of magnesium (oxide)				
Moles of magnesium (oxide)				
Mass of HCl solution				
Temp. change of HCl solution				
Heat of reaction, kJ/mol #				
Average heat of reaction				
Class average heat of reaction				

$q_{reaction} = - q_{HCl\ (aq)} = -$ (mass of HCl solution)(specific heat of HCl solution)(ΔT)

$\Delta H_{rxn} = (q_{reaction})$/(moles Mg or MgO)

(Assume the specific heat of 1 M HCl is the same as pure water and the heat capacity of the calorimeter is zero.)

1. Show your work for the calculation of the heat of reaction (ΔH_{rxn}) for Trial 1 (Mg and HCl).

2. Show your work for the calculation of the heat of reaction (ΔH_{rxn}) for Trial 1 (MgO and HCl).

OVER →

RESULTS and POST-LABORATORY QUESTIONS continued...

3. Applying Hess's law, use your experimental class averages for Equations 5 and 6 and the textbook value for Equation 7 (PRE-LABORATORY EXERCISE #4) to calculate a value for the heat of combustion of magnesium (Equation 4). Show all of your work.

4. Calculate the percent error between your experimental value for the heat of combustion of magnesium (POST-LABORATORY QUESTION #3) with the theoretical value for the standard enthalpy of formation of MgO from your textbook (ΔH_f^o, MgO).

5. Examine the experiment for possible sources of error. Discuss how you might improve the accuracy of the experiment.

Chemical Nomenclature II: Naming Basic Organic Compounds

Kristen Spotz

OBJECTIVES

- Recognize different representations of organic molecules.
- Build a variety of organic molecules using a model kit.
- Name simple organic molecules according to conventional guidelines.

INTRODUCTION

Due to the complex structures and compositions of organic compounds, many prominent thinkers in the 19th century believed that organic molecules contained a vital force. Friedrich Wohler who challenged this original belief claimed that "Organic chemistry is enough to drive one mad. It gives me the impression of a primeval tropical forest, full of the most remarkable things."

From the food we eat to the clothes we wear, organic compounds are all around us. Organic molecules have an indispensable role in research and industry. Our cosmetics, fuels, medicines and even our DNA contain organic compounds. Overall the vast majority of natural and man-made compounds are organic (over 16 million are known). Essentially life as we know it would be inconceivable without organic molecules. With such immense variety to deal with (Figure 1), how do chemists go about naming organic compounds?

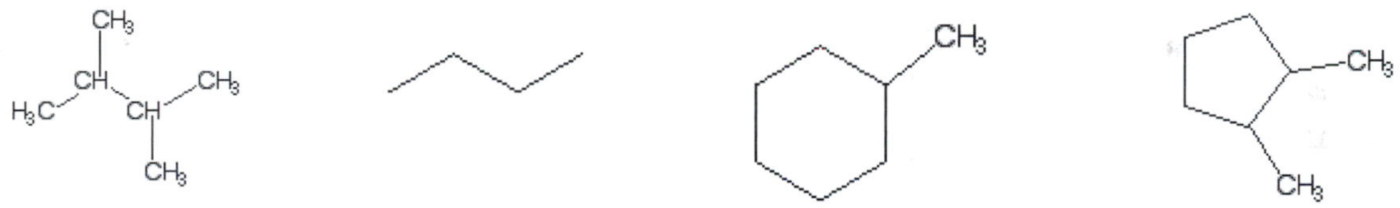

CH3 CH CH3 H3C CH CH3 CH3

Figure 1. Four examples illustrating the endless variety of hydrocarbons

BACKGROUND

The Variety of Organic Compounds

Organic chemistry is the study of compounds containing carbon, almost always bonded to itself and to hydrogen and sometimes a heteroatom, (for example, elements such as oxygen, nitrogen, sulfur, phosphorus, or a halogen). Carbon is unique in its ability to form four covalent bonds through different combinations of single, double, and triple bonds. In addition, carbon atoms form strong bonds with other carbon atoms creating long straight chains, branched chains, and rings of various sizes and shapes. With an endless variety of organic compounds, there is a need for a systematic method of naming so chemists can communicate. To address this need, the International Union of Pure and Applied Chemistry (IUPAC) devised a system of naming organic compounds which is now widely accepted.

Picturing Organic Molecules

It is often necessary for a scientist to convey not only information about which atoms are present in a molecule (the chemical formula), but also how the atoms are joined together in space. Recognizing the 3-D shape of organic molecules is crucial to understanding their physical properties and chemical reactivity. For this reason, chemists have various representations for picturing organic molecules. For example, the <u>molecular formula</u> of butane is C_4H_{10}. Various representations of butane are shown in the following figures.

The condensed formula (Figure 2a) groups the hydrogen atoms with the carbon atoms to which they are bound.

$CH_3CH_2CH_2CH_3$
or
$CH_3(CH_2)_2CH_3$

Figure 2a. Condensed formulas

The structural formula (Figure 2b) shows the individual atoms and bonds between them.

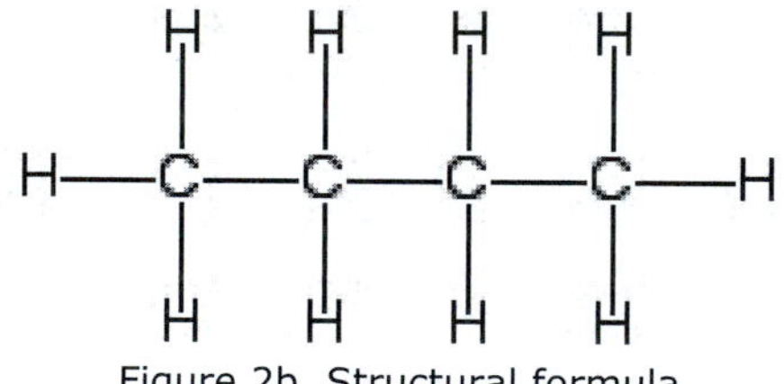

Figure 2b. Structural formula

The skeletal formula or line-bond formula (Figure 2c) consists of lines with carbon atoms at each intersection or at the end of a line. The hydrogens are then filled in to complete the octet of carbon.

Figure 2c. Skeletal diagram or bond-line formula

The perspective drawing (Figure 2d) displays the 3-D shape of the molecule by using solid lines to represent the bonds coming out of the page and dashed lines to represent the bonds going back into the page.

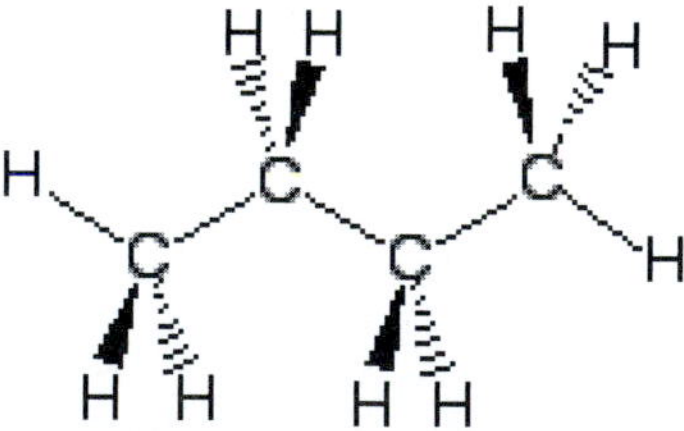

Figure 2d. Perspective drawing

The ball-and-stick model (Figure 2e) shows the atoms as spheres and bonds as sticks using accurate angles and relative sizes.

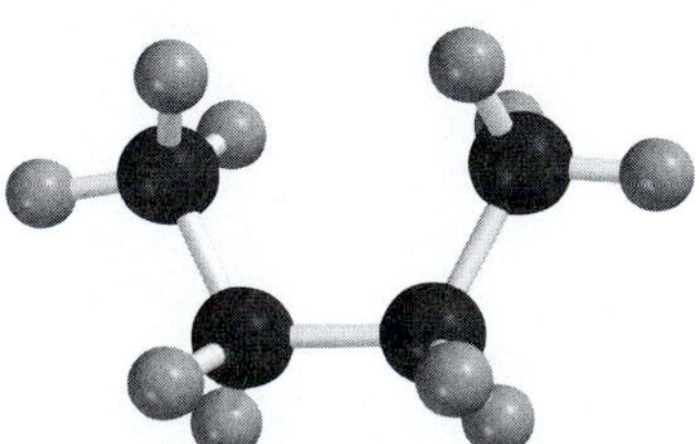
Figure 2e. Ball-and-stick model

The space-filling model (Figure 2f) is the more accurate representation of the molecule. However, this representation is not widely used due to its difficulty to draw.

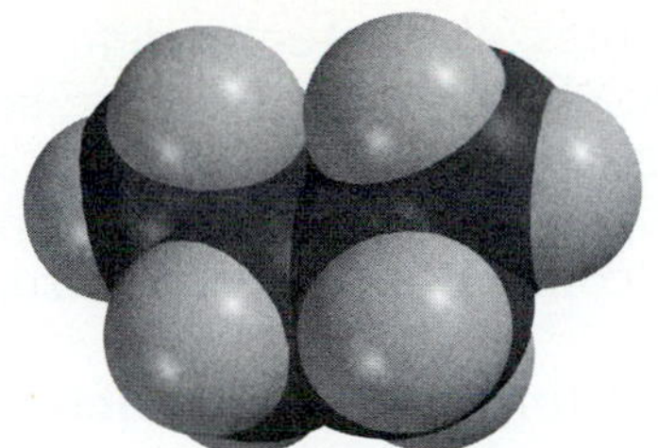

Figure 2f. Space-filling model

Nomenclature of Alkanes

All molecules in which each carbon is bonded to four other atoms are called alkanes. Because of this requirement, alkanes can have no double or triple bonds present (they are saturated). To name alkanes, identify the root word, which is dictated by the longest continuous carbon chain.

# of Carbon Atoms	Root Word
1	**meth-**
2	**eth-**
3	**prop-**
4	**but-**
5	**pent-**
6	**hex-**
7	**hept-**
8	**oct-**
9	**non-**
10	**dec-**

Table 1. Frequently used root words

All alkanes contain the root word and the suffix **-ane**. For example, a straight chain containing seven carbons and sixteen hydrogen atoms is heptane. Alkanes may also have a prefix attached to the root word. For example, if the chain consisting of seven carbons was in the form of a ring structure, the prefix **cyclo-** would be attached to the root word. Therefore, a seven-member carbon ring with fourteen hydrogen atoms is cycloheptane.

Isomers

One reason why there are so many different organic molecules is because the same number and kinds of atoms can be arranged in many different ways. Notice the two molecules (Figure 3a and b):

Figure 3a. Butane

Figure 3b. 2-methyl-propane

What is similar about the molecules and what is different? The molecule on the left has a linear chain of four carbon atoms with hydrogen atoms arranged on each carbon atom in order to complete the required octet. The molecule on the right is composed of the same number of hydrogen and carbon atoms, but the carbon atoms are arranged in an upside down T-shape. In other words, both molecules have the same molecular formula but have different bonds between atoms. In addition, the physical properties of the two molecules are different. For example, the molecule on the left boils at 0°C, whereas the molecule on the right boils at -12°C. Even though the two molecules have the same molecular formula the molecules are actually different. This phenomenon illustrates the concept of isomers. In organic chemistry, there are several types of isomers. However, this lab will focus on two types of isomers; structural and geometric isomers. In the example above the two molecules, butane and 2-methyl-propane, are structural isomers. Later in the lab, we will study the concept of geometric isomers.

Guidelines for Naming Branched Chain Alkanes

Example 1

- Locate the longest continuous carbon chain. This designates the molecule's parent chain and the root word of the molecule (Table 1). In Figure 4a the parent chain has four carbons so the root name is **but-**.

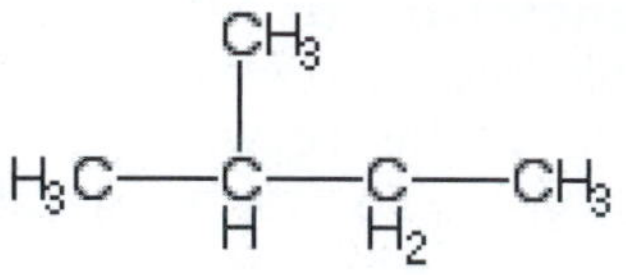

Figure 4a.

- Number the parent chain such that the branch (substituent) falls on the lowest carbon number. In Figure 4a, the parent chain is numbered so that the branch falls on carbon #2 instead of carbon #3.

- Because the branch is composed of only carbons and hydrogens (alkyl branch), use the same set of root words as in Table 1, but this time the suffix is **-yl.** In Figure 4a, the branched group, -CH_3, is a methyl group.

- Designate the location of the branched group by placing the number followed by a hyphen before the name of the branched group (2-methyl). The name of the branch serves as the suffix of the parent chain. In Figure 4a, the compound is called 2-methylbutane.

Example 2

- If there is more then one particular substituent (Figure 4b), designate the position of every group and alphabetize the substituents.

3-ethyl-2-methylhexane

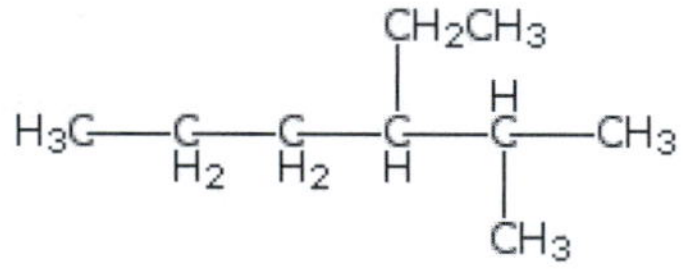

Figure 4b.

Example 3

- If there is more than one of the same branch on a chain (Figure 4c), designate the positions of the substituent and name the branch with the additional prefix indicating how many times it occurs in the molecule: **di -**(2), **tri-** (3), **tetra-** (4).

2,2-dimethylbutane

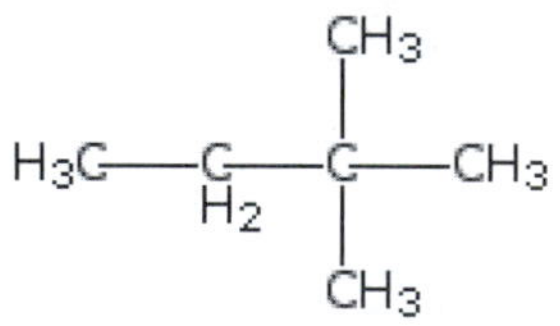

Figure 4c.

Nomenclature of Halide branches

Alkyl halides are alkanes with halogen atom substituents. The root chain is named the same way as we have seen for regular carbon branches. Halogen branches are indicated with the following prefixes added to the name of the parent chain:

	-F	-Cl	-Br	-I
Branch Name	fluoro-	chloro-	bromo-	iodo-

Assign appropriate prefixes (di-, tri-, tetra-) if there is more than one type of halogen present in the molecule. The branches are listed in alphabetical order when naming the molecule.

Naming Alkenes

Alkenes are unsaturated hydrocarbons that contain double bonds. The root words for alkenes denoting the longest continuous carbon chain are the same as alkanes. The only difference is the name ends in **-ene** instead of -ane. In addition, the position of the double bond is indicated by a number, which designates the first carbon atom involved in the double bond. This number is followed by a hyphen and placed in front of the parent name. Our example below is: 2-methyl-3-octene.

Notice that the longest continuous carbon chain contains 8 carbons and the parent chain is numbered so that the double bond begins at the lowest number. Beginning on the right side of the molecule, the methyl group is on second carbon and the double bond occurs between carbons 3 and 4.

Actually, there is more than one name for the above compound because with alkenes we must deal with a new type of isomer, the geometric isomer. Geometric isomers have the same molecular formula, but because of their inability to rotate about a certain part of the molecule, there are different possible spatial arrangements of the atoms. To help illustrate this concept, construct the following pair of molecules:

Are the above molecules the same? By rotating the molecule on the left about the C-C single bond, we see that both molecules are actually the same. We expect them to have the same properties and we give them the same name (1,2-dibromoethane). Now, look at the alkenes below:

The molecules are restricted from free rotation about the double bond resulting in two distinct molecules. To differentiate between the molecules, chemists use the term **cis-** meaning the substituents are on the same side of the double bond and **trans-** meaning the substituents are on opposite sides of the double bond. The molecule on the left is trans-1,2-dibromoethene and the molecule on the right is cis-1,2-dibromoethene.

Nomenclature of Alkynes

Alkynes are unsaturated hydrocarbons containing a carbon-carbon triple bond. Alkynes are named by identifying the longest continuous carbon chain containing the triple bond and adding the suffix **–yne** onto the root word. As with alkenes, the position of the triple bond is identified.

OVERVIEW

In this organic tutorial, you will practice building and naming alkanes (both straight and branched chains), alkenes, alkynes, alkyl halides and some basic ring structures.

PROCEDURE

Materials Used
Organic model building kit

As you complete the following procedure, following along and answer the questions on the **RESULTS and POST-LABORATORY QUESTIONS** page.

Part A. Nomenclature of Alkanes

1. Construct the three isomers of **pentane** using your model set. Be careful that some of your "isomers" aren't really just the same molecule that has been rotated either in space or about a C-C single bond.

2. Using your model kit, build the five isomers of **hexane**.

Part B. Nomenclature Alkyl Halides

3. Construct the following alkyl halides with your model set.

a)

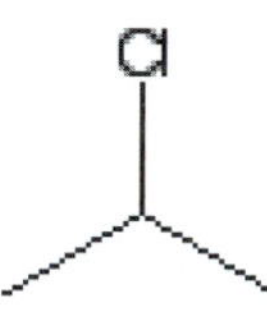

b)

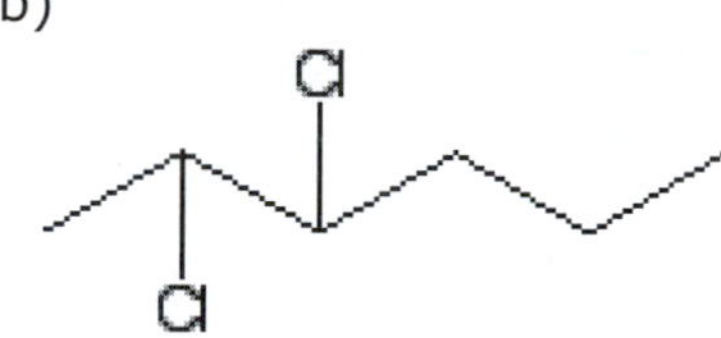

c)

Br

Part C. Nomenclature of Alkenes

4. With your model kit, construct the two geometric isomers of **2-butene**.

5. Construct the two geometric isomers of **1,2-dimethylcyclopropane**.

Part D. Nomenclature of Alkynes

6. Build an alkyne of your choice with four carbons and the required number of hydrogens.

Chemical Nomenclature II: Naming Basic Organic Compounds

Name:	Lab Instructor:
Date:	Lab Section:

PRE-LABORATORY EXCERCISES

1. Define the underlined words in the BACKGROUND section.

2. Draw the Lewis Dot Structure for methane, CH_4. According to VSEPR theory, what is the molecular geometry of methane? What are the predicted bond angles?

3. Circle and give the number of carbons in the longest continuous chain in each of the following molecules.

a) b) 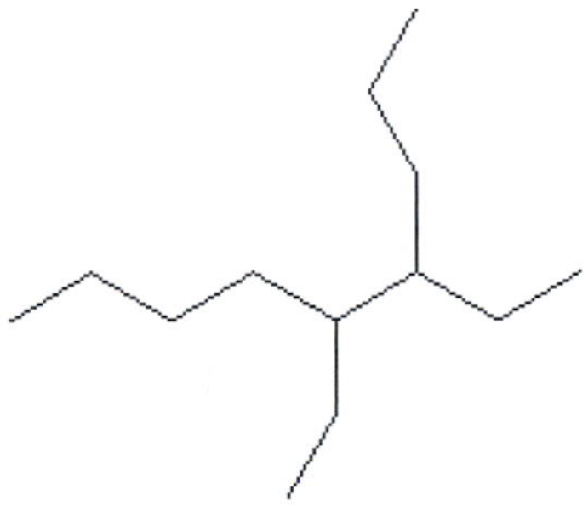 c)

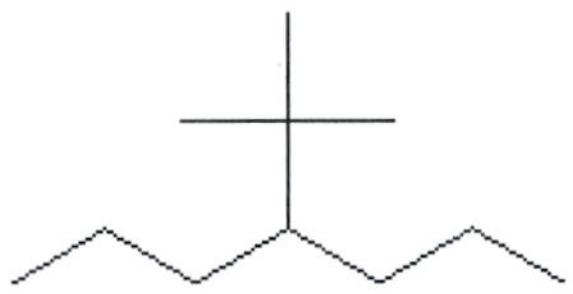

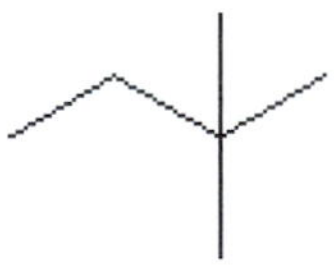

OVER →

PRE-LABORATORY EXCERCISES continued...

4. Redraw each of the following representations as the corresponding structural formula.

a)

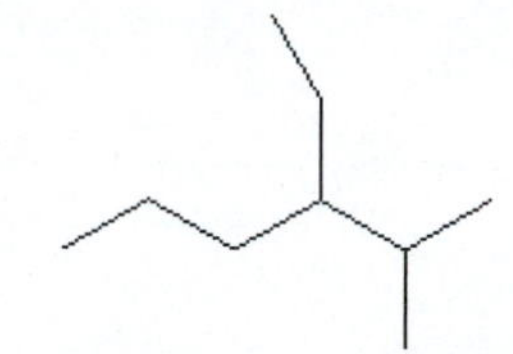

b)

c) $(CH_3)_3CC(CH_3)_2CH_2CH_2CH_3$

Chemical Nomenclature II: Naming Basic Organic Compounds

Name:	Lab Instructor:
Date:	Lab Section:

RESULTS and POST-LABORATORY QUESTIONS

Part A. Nomenclature of Alkanes

Draw the **structural formulas** for the three isomers of **pentane**. Name the isomers.

Structural formula:			
Name:			

Draw the **skeletal formulas** of the five isomers of **hexane**. Name the isomers.

Skeletal formula:			
Name:			

Skeletal formula:		
Name:		

OVER →

Part B. Nomenclature Alkyl Halides

Draw the **structural formulas** and name each of the following molecules:

a) b) c)

Cl Cl Cl Br

	a)	b)	c)
Structural formula:			
Name:			

Part C. Nomenclature of Alkenes

Draw the **structural formula** and name the two geometric isomers of 2-butene.

Structural formula:		
Name:		

To prove that the two geometric isomers of 2-butene are really two different compounds, find their boiling points in your textbook or on the internet.

Continued on next page →

Chemical Nomenclature II: Naming Basic Organic Compounds

Name:	Lab Instructor:
Date:	Lab Section:

RESULTS and POST-LABORATORY QUESTIONS continued...

Part C. Nomenclature of Alkenes continued...

Draw the **structural formula** and name the two geometric isomers of 1,2-dimethylcylopropane. Name the isomers.

Structural formula:		
Name:		

Part D. Nomenclature of Alkynes

Draw the **structural formula** of the alkyne you built.

Is geometric isomerization possible about a triple bond? Explain.

Supplemental Problems

1. Name the molecules in Figure 1 in the INTRODUCTION.

OVER →

RESULTS and POST-LABORATORY QUESTIONS continued...

Supplemental Problems continued...

2. What type of formula is indicated in each of the following? Redraw each compound using the **structural formula**. Name each compound.

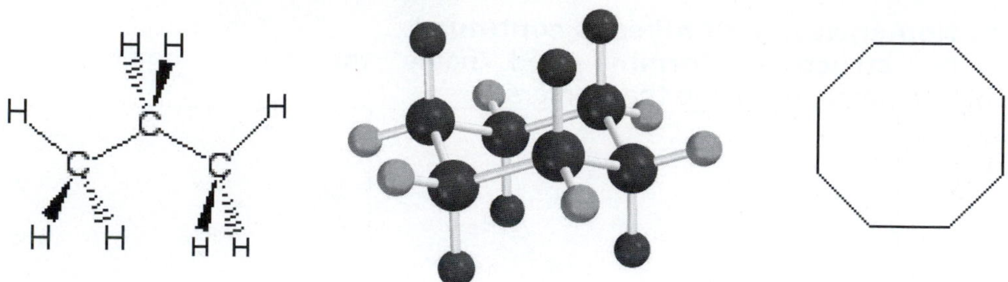

Formula Type:			
Structural Formula:			
Name:			

3. Draw the **structural formula** for the three isomers of C_3H_4.

4. Name and draw the **skeletal formula** for any molecule that has six carbons, one double bond, and a halogen.

Introduction to Organic Analysis: Infrared Spectroscopy

Kristen Spotz

OBJECTIVES

- Identify specific functional groups within an organic molecule.
- Review VSEPR theory in the context of organic molecules.
- Discuss the basic principles and applications of infrared spectroscopy.
- Use the results of IR spectroscopy to practice identification of functional groups and basic structure determination of organic molecules.

BACKGROUND

This summer you are working as a forensic science intern in a crime scene laboratory. On your first day, you help gather evidence from a jewelry theft that was the result of a home invasion. A week later, a possible suspect's car and home are searched and a crowbar is found that could have been used to pry open the door to the victim's house. It is your job as a forensic science intern to match the crowbar with evidence found at the crime scene. A small amount of paint was found on the crowbar. Does it match the paint on the back door? You run a method of chemical analysis, known as infrared spectroscopy, because it is one of the few tests that can be used for positive identification of paint samples. From your results (Figure 1), you are able to help convict a jewel thief and prove yourself to be a competent forensic scientist off to a great career.

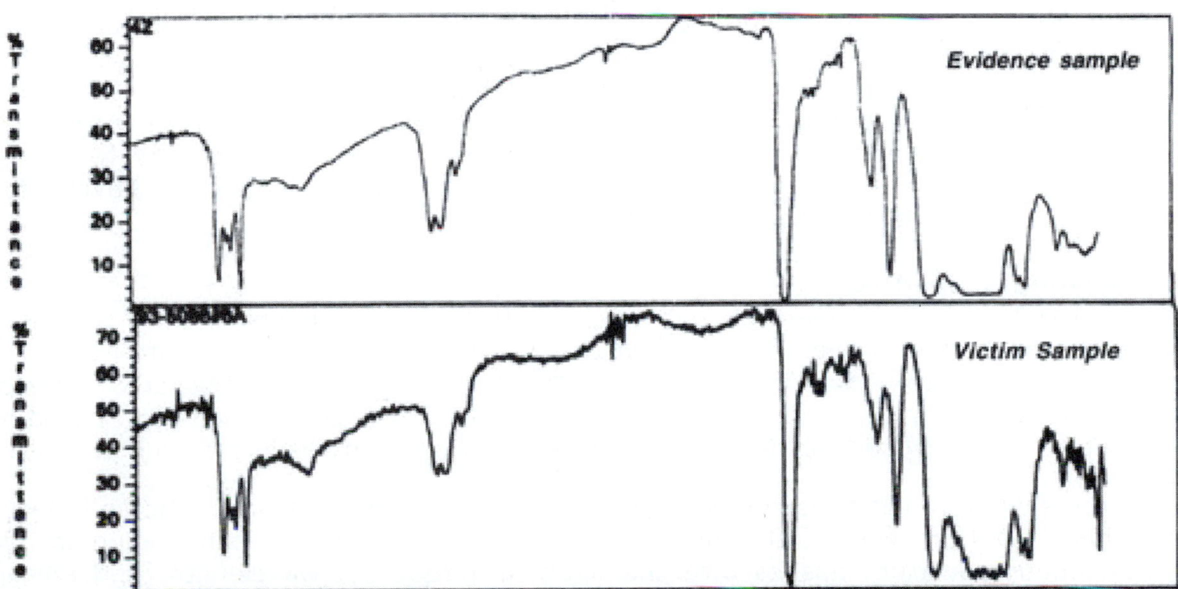

Figure 1. Paint samples compared by infrared spectrometry. In the figure, a sample of evidence paint from the crowbar (top) is compared with a sample of paint from the back door.

Functional Groups

One of the most useful concepts in chemistry is organization of the vast number of organic molecules according to their functional groups. Any bonded group of atoms, other than the carbon-carbon bond or the carbon-hydrogen bond, constitutes a functional group. Table 1 is a list of common functional groups that you need to familiarize yourself with.

Table 1. Common functional groups

Functional Group	Compound Type
—C=C—	Alkene
—C≡C—	Alkyne
—OH	Alcohol
—C—C(=O)—C—	Ketone
—C(=O)—H	Aldehyde

Functional Group	Compound Type
—C≡N	Nitrile
$—NH_2$	Amine
—C—O—C—	Ether
—C(=O)—OH	Carboxylic acid

Functional groups are where the chemical reactions typically take place on a molecule. Each functional group reacts in a characteristic way regardless of what other carbon framework it is attached to. To predict how a molecule behaves during a chemical reaction, we narrow our focus to the functional group because the distribution of electron density in the functional group affects chemical reactivity. Suppose a drug was synthesized at a pharmaceutical company and the chemist needs to know how the drug will interact with other chemicals. The first step would be to identify the functional group(s) in the molecule. Fortunately, chemists have a technique known as infrared (IR) spectroscopy to identify functional groups. To understand how IR spectroscopy works, we need to have a firm understanding of the electromagnetic spectrum.

Electromagnetic Spectrum

In the late 1600's, Sir Issac Newton was experimenting with the nature of light. By shining a narrow beam of light through a glass prism, he was able to separate the white light into a rainbow of colors. Later it was found that there are regions of radiation that are not visible to the human eye. For example, the region just outside of the visible region (now called the IR) can be detected because it raises the temperature of a thermometer placed in its path.

All electromagnetic radiation is composed of oscillating perpendicular magnetic and electric fields that travel in waves at the speed of light. However, each region of the electromagnetic spectrum has its own characteristic frequency, wavelength, and energy. For example, infrared radiation has a higher frequency, shorter wavelength and higher energy than radio waves. It is these differences that result in each region of the electromagnetic spectrum having a unique ability to interact with matter. Different regions of the spectrum cause specific molecular responses. For example, high-energy radiation, such as ultraviolet light, has enough energy to excite an electron from one molecular orbital to another. Less energetic radiation, such as infrared radiation, causes changes in the vibrational energy of a molecule. On the lowest energy end of the spectrum, radio waves have only enough energy to excite molecules to rotate. This is the cornerstone of spectroscopy, which is a set of analytical methods that seek to gain information about molecules by observing how the molecules interact with various regions of the electromagnetic radiation.

Infrared Spectroscopy

Infrared (IR) spectroscopy was one of the first tools used by chemists to determine the chemical structure of an unknown molecule. Even though more advanced instruments have since been designed, infrared spectroscopy is still widely used today due to its ability to identify what functional groups are present in a molecule or as proof that two compounds are identical. An IR spectrometer selects a particular wavelength of the IR and passes the light (radiation) through a sample. If the light does not have the required energy, the light does not interact with the sample and passes through to the detector. However, at wavelengths characteristic of each molecule, the sample will absorb the light. The spectrometer detector measures the extent to which the sample absorbs the light.

As mentioned above, the basis of infrared spectroscopy lies in the molecular vibrations of the molecule. The atoms in a molecule can be thought of as balls attached by springs. If the "balls" are stretched apart, they pull back with a restoring force, causing the ball and spring to vibrate. Unlike a ball and spring, however, a molecule cannot vibrate at any given frequency. Molecules only vibrate at specific frequencies corresponding to quantized energy levels. This concept is similar to the quantized energy levels of an electron in an atom. Thus, when a molecule absorbs a photon of light in the required range (IR range: 2.5×10^{-6} m to 16×10^{-6} m), the molecule is excited from its lowest, or ground, vibrational state to a higher energy, excited vibrational state with an increased vibrational frequency.

Each molecule has different kinds of molecular vibrations including stretching (Figure 2) and bending patterns (Figure 3). Most importantly, each functional group has a characteristic vibrational frequency that is relatively independent of its molecular environment. For example, an oxygen-hydrogen bond in an alcohol has roughly the same vibrational frequency regardless of what it is attached to.

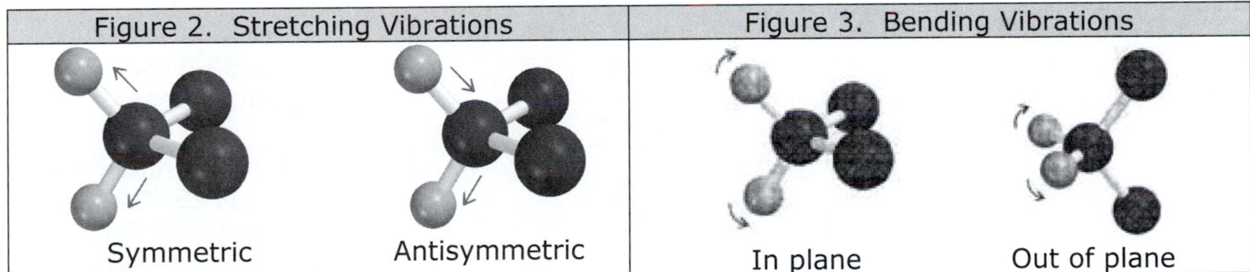

A typical spectrum from infrared spectroscopy appears as a series of absorption bands ranging in intensity and shape (ex. width, sharpness). The x-axis designates the frequency of light expressed in wavenumbers (inverse centimeters, $1/\lambda$). Wavenumbers are a more convenient convention than frequency because wavenumbers are directly proportional to energy. The y-axis of the spectra designates percent transmittance (%T).

When using infrared (IR) spectroscopy for structure determination, absorption bands in the range of 1600-4000 cm^{-1} are most useful because this range is where the vibrational frequencies of functional groups are located. In addition, bands between 1300-625 cm^{-1} fall in the "fingerprint" region because no two compounds have identical spectra in this range of wavenumbers. Tables 2 and 3 list the cm^{-1} associated with some common organic functional groups and various types of C-H bonds.

Table 2. Infrared Absorption of Hydrocarbons

Hybridization on Carbon	Category of Methyl Group	Wavenumber (cm^{-1})
sp^3 C-H	—CH_3 (terminal) —C(H_2)— (internal)	1470-1370 and 2970-2850 2940-2850 and 1470-1450
sp^2 C-H	=C(H)(H) (terminal) (H)(—)C= (internal)	3090-3070 3030-2980
sp C-H	—C≡C—H (terminal)	3300-3250 and 680-650

Table 3. Infrared Absorption of Some Common Functional Groups

Functional Group Name	Representation	Wavenumber (cm^{-1})
Alcohols	—O—H	3600-3200 (broad) 1200-1025 only C-OH
Aldehydes	—C(=O)—H	1440-1320 and 1750-1710
Alkenes	—C=C—	1680-1620
Alkynes	R—C≡C—H	2250-2100 (C≡C)
Amines	—N(H)—H	3450-3350 (pair of peaks)
Carboxylic acids	—C(=O)—OH	3600-2500 broad O-H 1700-1725 (C=O) 960-910 (C-OH) 700-590 (O-C=O)
Ethers	—C—O—C—	1200-1025
Ketones	—C—C(=O)—C—	1725-1705

Analysis of IR Spectra

Example 1: hexane

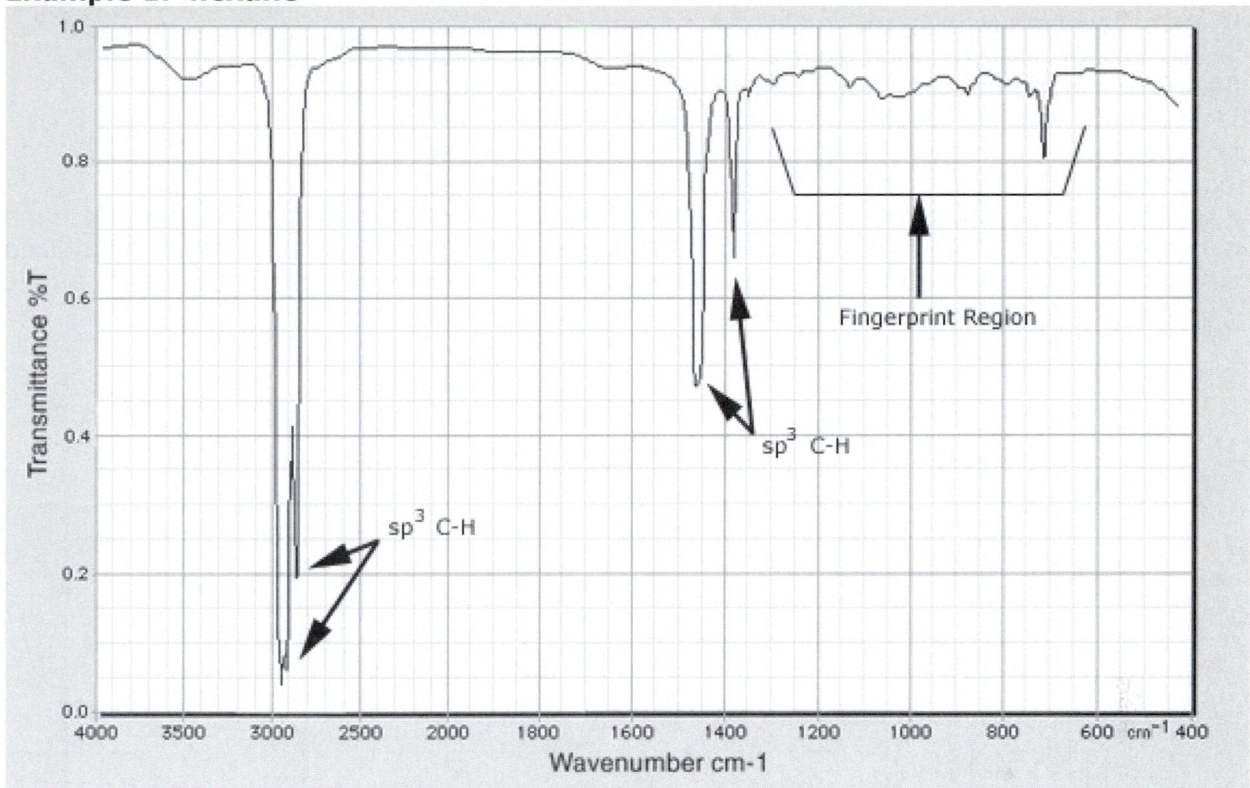

Figure 4. Sample spectrum of hexane.

To begin analyzing the various regions of absorption on the spectrum for hexane (Figure 4), first the hybridization of each carbon atom must be evaluated. All the carbon atoms in hexane are sp^3 hybridized (Figure 5). Referring to Table 2, there are two categories of sp^3 hybridized carbons, either CH_2 or CH_3 groups. In hexane, there are 4 CH_2 groups and 2 CH_3 groups. Therefore, we expect absorption bands in the regions of 2970-2850 cm^{-1} and 1470-1370 cm^{-1} for the CH_3 groups and absorption bands in the regions of 2940-2850 cm^{-1} and 1470 – 1450 cm^{-1} for the CH_2 groups. The actual absorption bands in Figure 4 are located, as expected, at 2960 cm^{-1}, 2872 cm^{-1}, 1460 cm^{-1} and 1375 cm^{-1}. Notice that the regions of absorption for the CH_2 groups and CH_3 groups overlap. Unless those absorption regions are highly resolved, it is difficult to differentiate between the two groups of sp^3 hybridized carbons. However, for a simple spectrum such as hexane, a chemist can gather that the molecule contains only CH_2 and CH_3 groups connected by single bonds. Then, if the molecular weight of the molecule is known, the chemical formula and the structure can be derived.

$H_3C-CH_2-CH_2-CH_2-CH_2-CH_3$

Figure 5. Hexane

Example 2: 1-hexene

How would you expect the IR spectrum of 1-hexene to appear? To begin, identify any functional groups and the hybridization of each carbon atom in the molecule. 1-hexene contains a carbon-carbon double bond so the molecule should have an absorbance band between 1680-1620 cm^{-1}. The molecule also has two sp^2 hybridized carbon atoms. One of these carbon atoms has two terminal hydrogens, $=CH_2$, so absorbance bands should appear in the region of 3090-3070 cm^{-1}. The other sp^2 hybridized carbon atom has one hydrogen (HC=C) so it should exhibit an absorbance band in the region of 3030-2980 cm^{-1}. Finally, 1-hexene has both types of sp^3 hybridized carbons, so absorbance bands are expected in those regions. Figure 7 shows a sample spectrum of 1-hexene.

Figure 6. 1-hexene

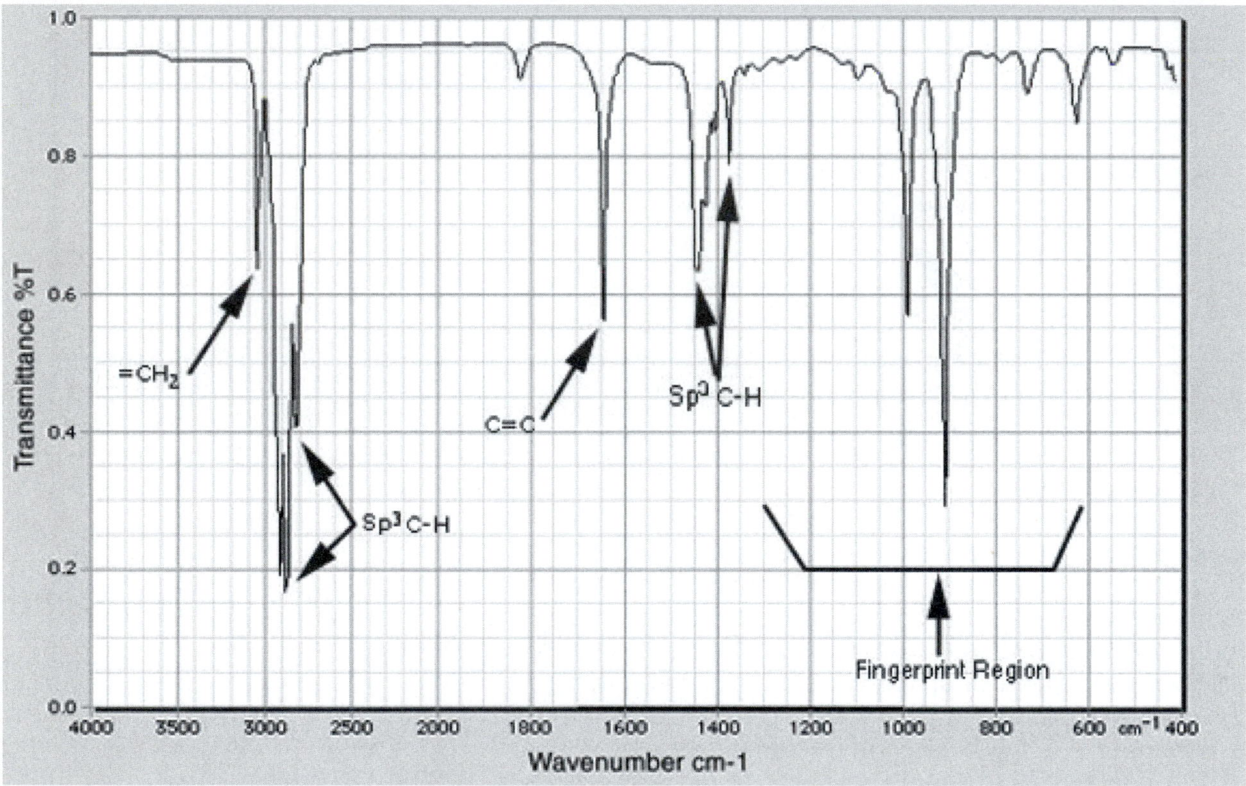

Figure 7. Sample spectrum of 1-hexene.

As predicted there is a characteristic absorbance at 3080 cm^{-1} due to the stretching vibration of the two hydrogens attached to the sp^2 hybridized carbon on the double bond. There is also an absorbance at 1642 cm^{-1}, which is in the 1680-1620 cm^{-1} diagnostic region for a carbon-carbon double bond. In addition, by comparing Figure 7 and Figure 4 we see there are also bands due to the carbon-hydrogen bonds on the sp^2 and sp^3 hybridized carbon.

Example 3: Unknown

Now, suppose you were given the spectrum in Figure 8 and assigned the task of identifying the functional group (s) and a preliminary structure for the molecule.

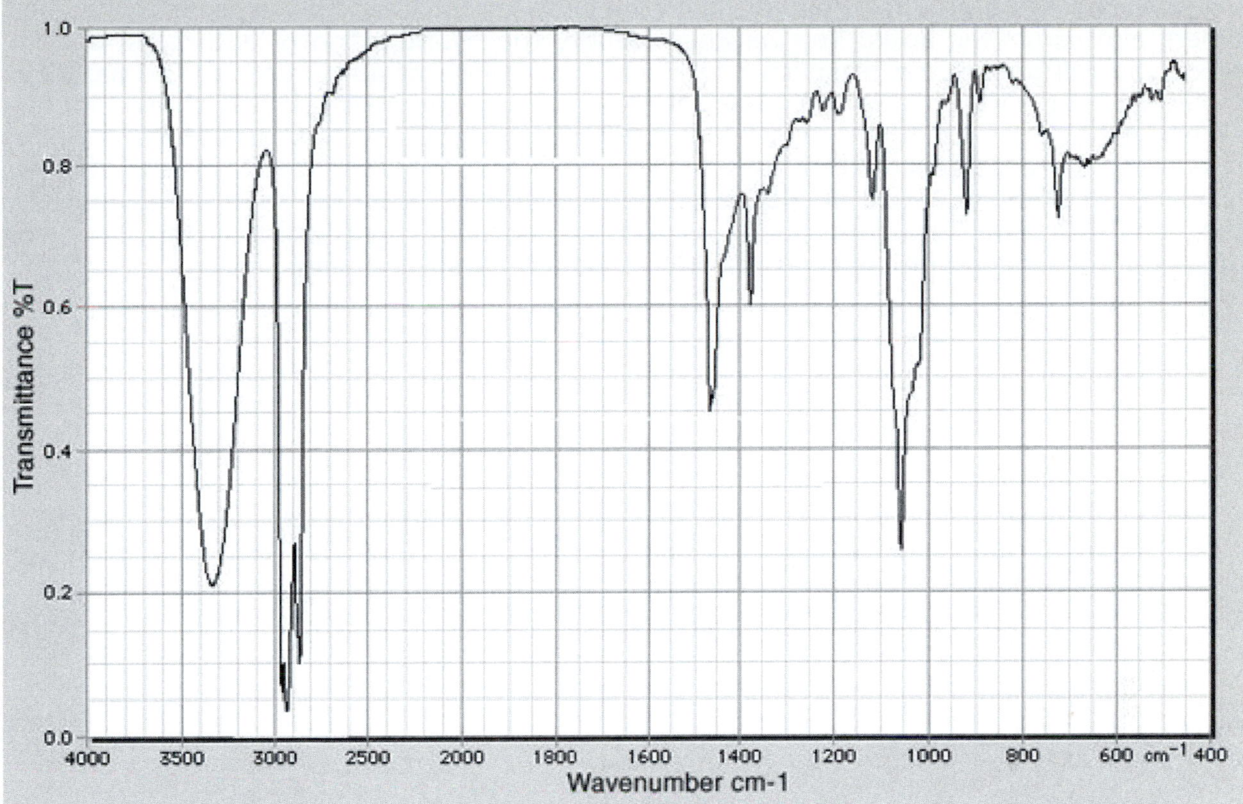

Figure 8. Sample spectrum of hexanol

You would begin by reading the spectrum from left to right, looking for characteristic regions of absorption. You notice an absorption at 3334 cm^{-1} which is diagnostic of either an alcohol or an amine. However, the band is broad which indicates the presence of an alcohol. Next, there are absorption bands located between 2960-2861 cm^{-1} and 1470-1380 cm^{-1}, which are due to CH_2 and CH_3 groups as we saw in the spectrum for hexane and 1-hexene. Finally, there is an absorbance at 1058 cm^{-1}, which is due to the vibration between a carbon atom and the oxygen atom from an alcohol. We would conclude the molecule in question is an alkyl chain with an alcohol functional group. If mass spectroscopy was run on the molecule and the molecular weight is determined to be 102.2 g/mol our unknown must be one of the isomers of $C_6H_{13}OH$. A reasonable choice would be hexanol (Figure 9). The confirmation of the unknown as hexanol could be made by comparing the fingerprint region of our spectrum to a spectrum known to be hexanol.

$H_3C-CH_2-CH_2-CH_2-CH_2-CH_2-OH$

Figure 9. Hexanol

OVERVIEW

In Part A of this experiment, you will be building various organic molecules containing the following functional groups: double and triple bonds, alcohols, amines, aldehydes, ketones, ethers and carboxylic acids. You will then identify the hybridization of each carbon atom. Next, you will predict the IR spectra by identifying regions of absorption for the particular molecule. In Part B, you will play detective solving the mysteries in three different scenarios. In Scenario 1, the concepts of qualitative analysis and structural isomers will be reviewed. Infrared spectroscopy will be used as a tool to differentiate between the two isomers. In Scenario 2, two IR spectra will be compared to locate the spectral region where the nitrile functional group is found. Finally, in Scenario 3, the chemical formula of an organic compound will be matched with its appropriate IR spectral data.

PROCEDURE

Materials Used
Organic model building kit

As you complete the following procedure, follow along and answer the questions on the **RESULTS and POST-LABORATORY QUESTIONS** page.

Part A. Functional Groups

A) CH_3CH_2CCH B) $CH_3(CH_2)_2O(CH_2)_2CH_3$ C) $CH_3(CH_2)_4CH_2OH$ D) $CH_3CH_2NH_2$
E) CH_3COH F) $CH_3CH_2CO(CH_2)_2CH_3$ G) CH_3CH_2COOH

Construct molecules A - G using your model set and draw the Lewis dot structure for each molecule in your laboratory notebook.

Part B. Detective Work

Scenario 1: You are a new hire in the chemical stockroom at your school. The stockroom has the chemicals A - G listed in Part A. One day, you find a small, unlabeled bottle on the shelf. You know it must be one of the chemicals above, but which is it? You run three analysis techniques on the compound and discover the following:

Analysis #1: Elemental analysis indicates the compound contains 15.66% mass by oxygen, 70.52% mass by carbon, and 13.82% mass by hydrogen.
Analysis #2: Mass spectrometry indicates the unknown compound has a molecular weight of 102.2 g/mol.
Analysis #3: Infrared spectroscopy indicates the unknown compound has the following spectra:

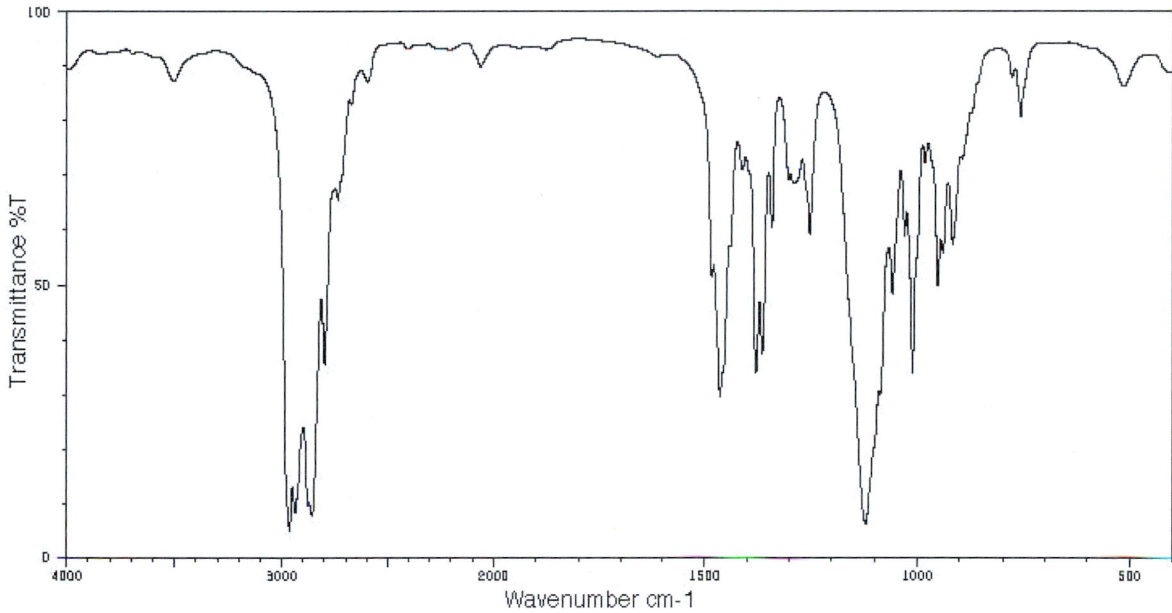

Scenario 2: The laboratory manual authors forgot to include in Table 3 the wavenumbers where nitriles exhibit characteristic absorption bands. Compare the IR spectra of hexane (Figure 4) and heptyl cyanide (Figure 10 and 11) to decipher this mystery.

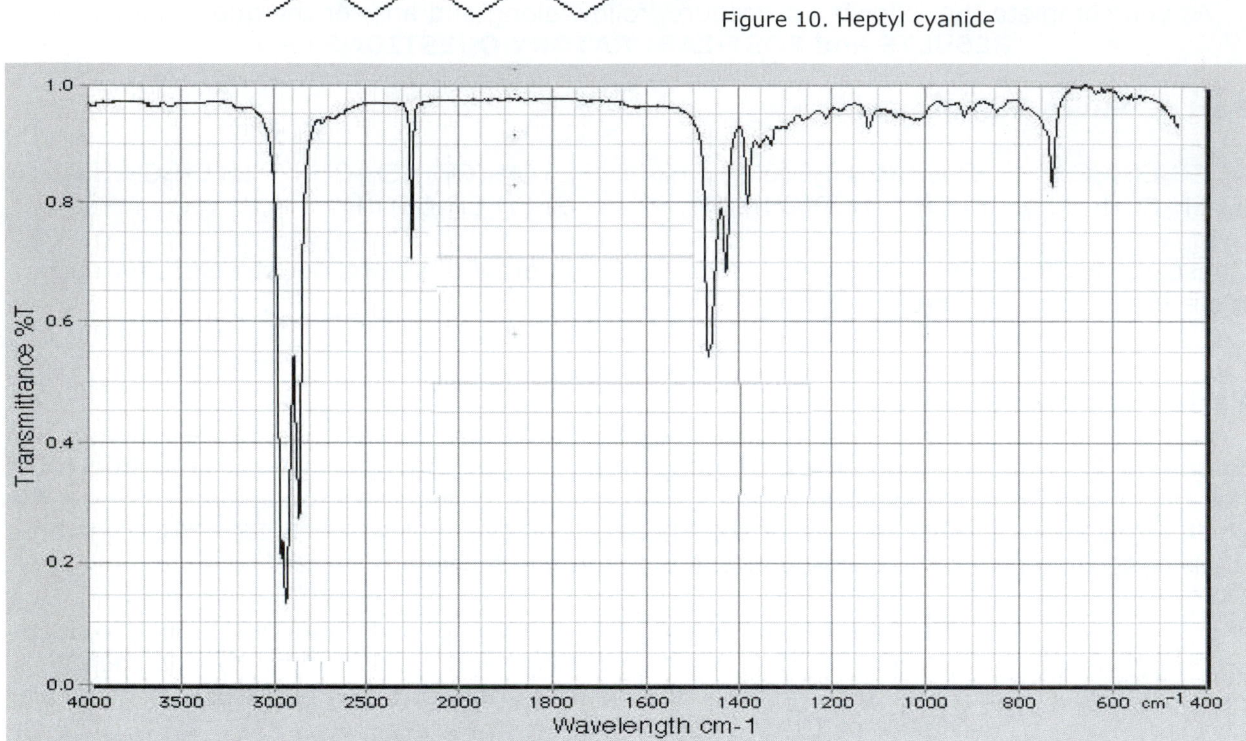

Figure 10. Heptyl cyanide

Figure 11. The IR spectra of Heptyl cyanide.

Scenario 3: Hidden in the closet in the back of the stockroom, you find six unlabeled bottles of organic compounds. Solving this situation is easier than Scenario 1, however, because you also find that the labels from the six different bottles have fallen off onto the floor. The formulas that appear on the labels are as follows:

H) CHOOH
I) $CH_3(CH_2)_4CCH$
J) $H_2NCH_2CH_2OH$
K) CH_2OHCH_2OH
L) $CH_3(CH_2)_3COCH_3$
M) $C_6H_{13}NH_2$

You decide to take a small sample from each bottle and run an IR spectrum. The results (Spectra 1 - 6 on the following 3 pages) will help you identify the contents of each bottle.

Construct molecules H - M using your model set and draw the Lewis dot structure for each molecule in your laboratory notebook.

Spectrum 1:

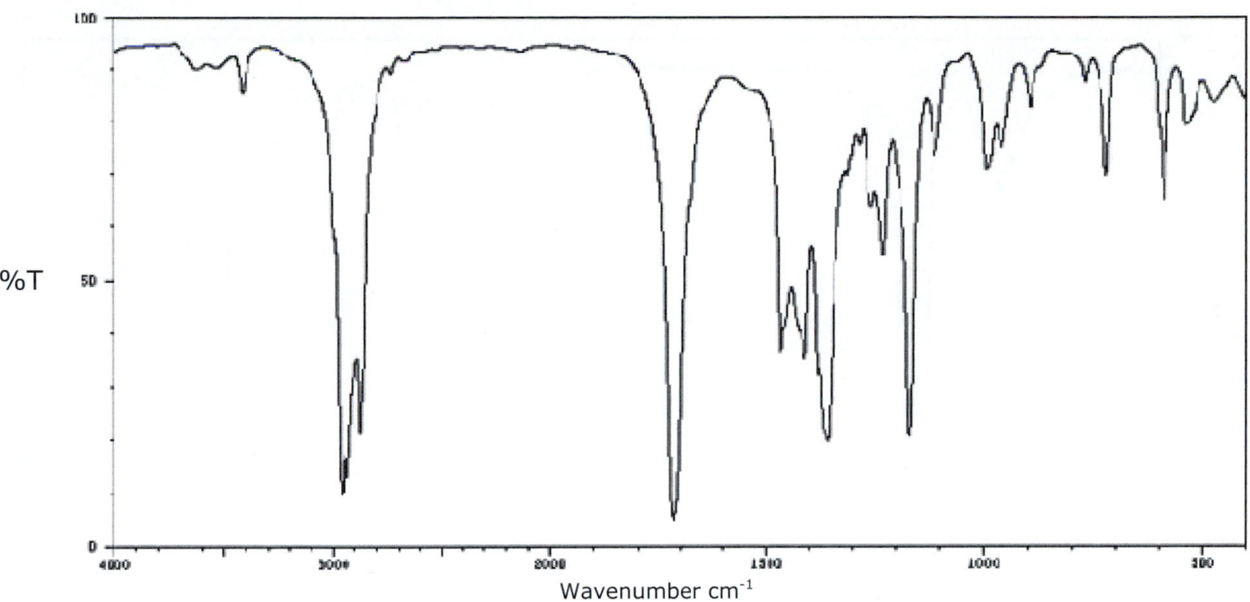

Spectrum 2:

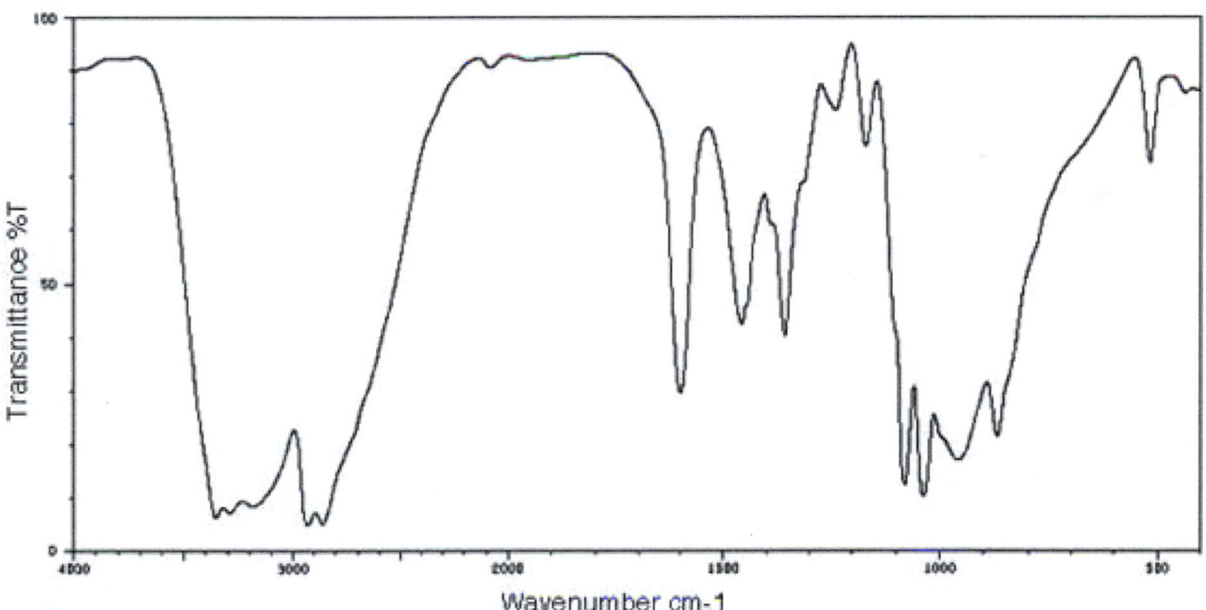

Spectrum 3:

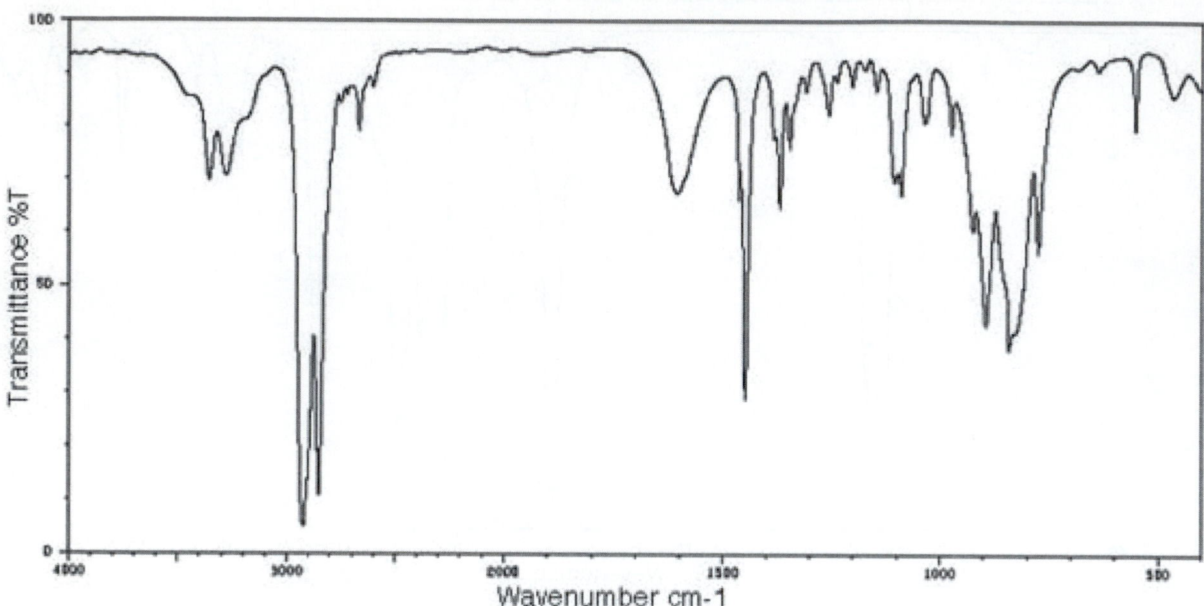

Spectrum 4:

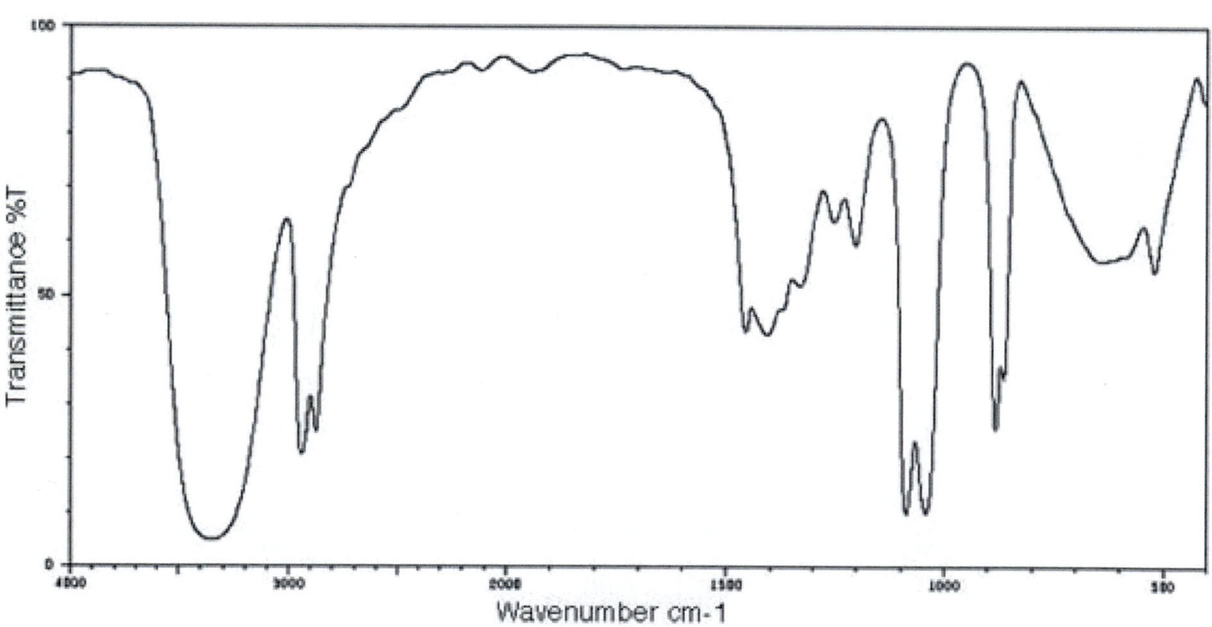

Spectrum 5:

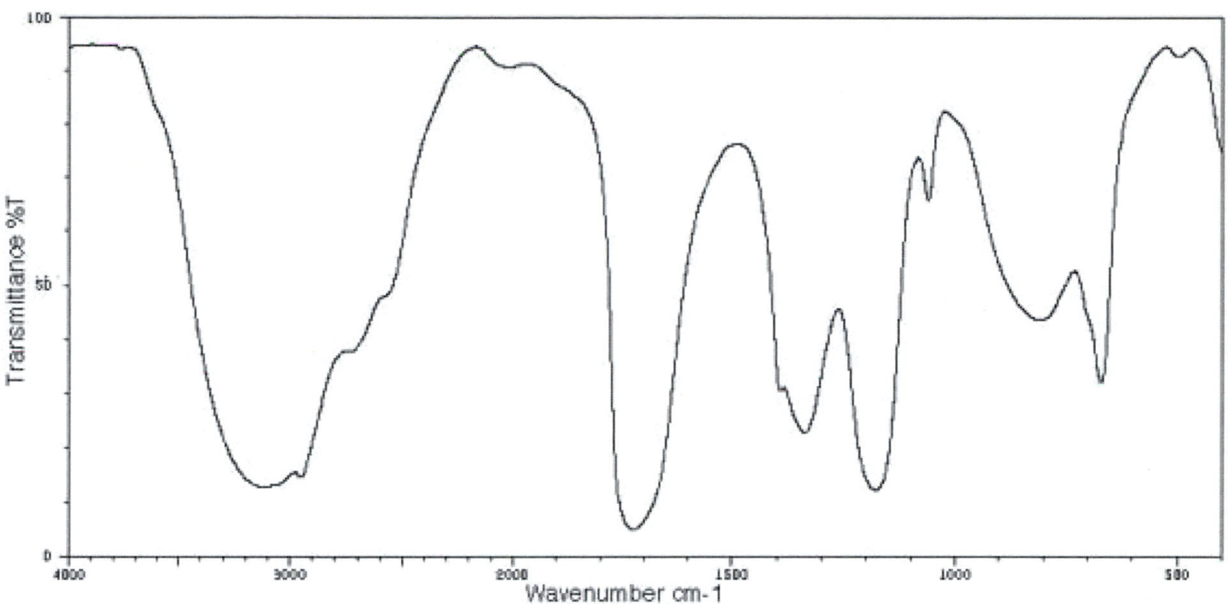

Spectrum 6:

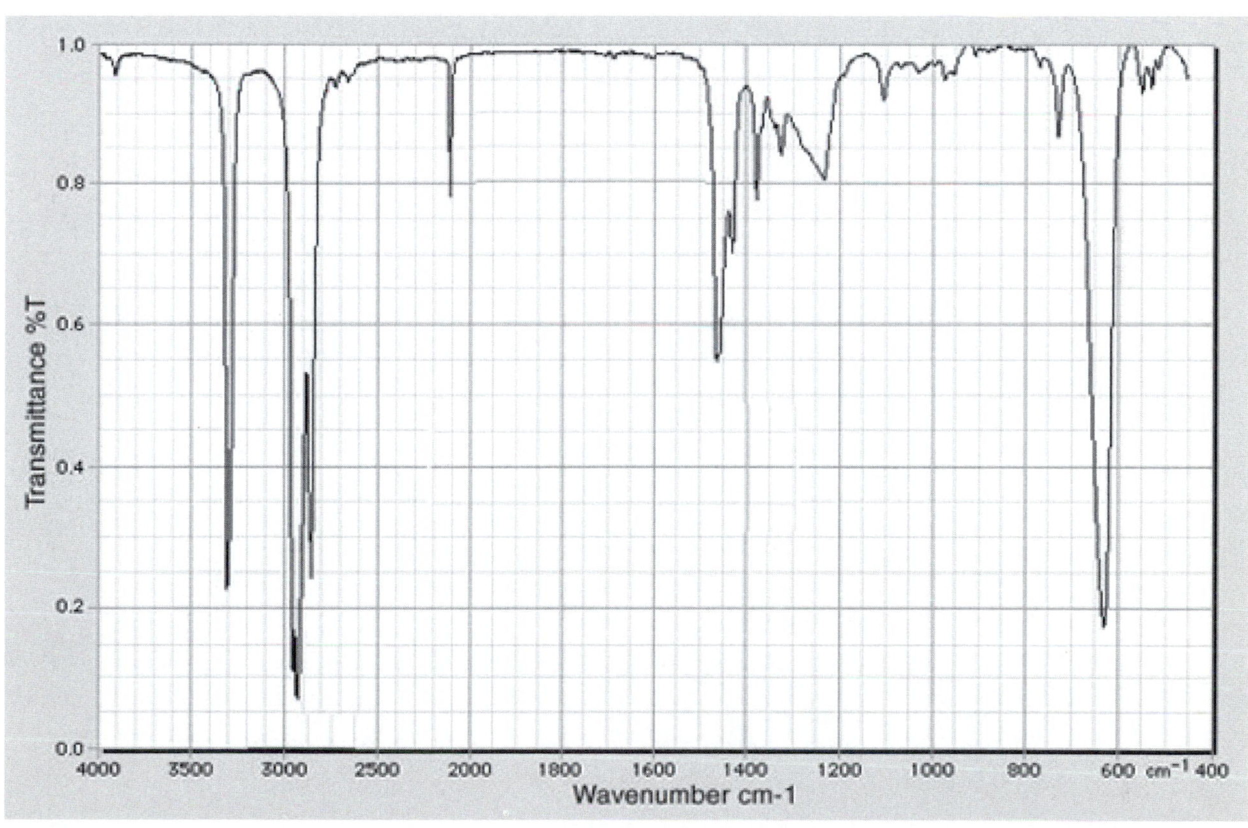

Introduction to Organic Analysis: Infrared Spectroscopy

Name:	Lab Instructor:
Date:	Lab Section:

PRE-LABORATORY EXERCISES

1. Circle and name the functional group(s) in the following molecules. Identify the hybridization of each carbon atom.

$H_3C—C≡N$

$H_3C—CH=CH—CH_2—O—CH_3$

$CH_3C(=O)CH_2CH_3$

(cyclohexa-1,3-dienyl)$—CH(OH)—CH_2—NH—CH_3$

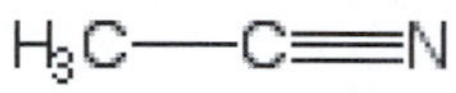

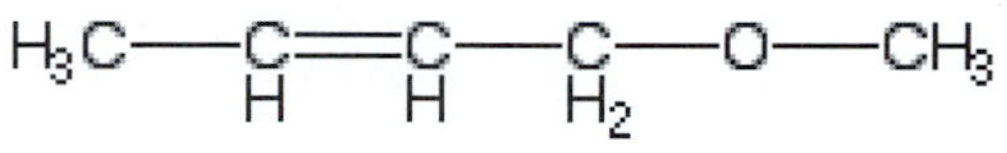

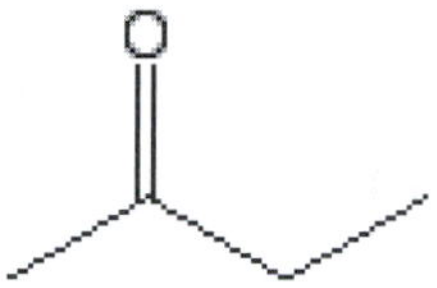

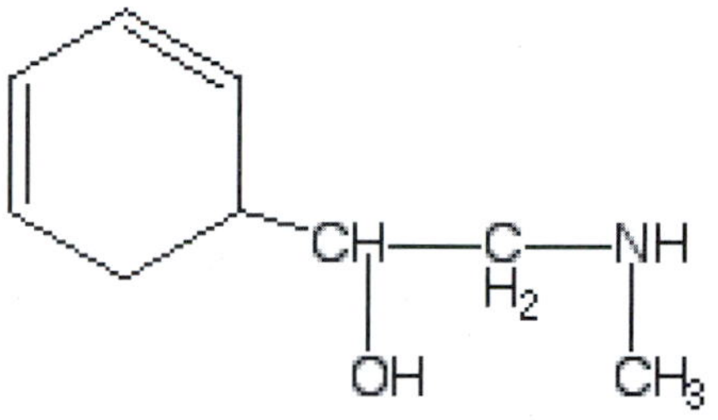

2. Explain how you would differentiate between hexane and hexyne by infrared spectroscopy. Your explanation should include a drawing of their structural formulas. Label the hybridization of each carbon atom and the expected absorbance region(s) for each molecule.

OVER →

PRE-LABORATORY EXERCISES continued...

3. Draw the possible alcohols with the formula $C_4H_{10}O$. What principle absorption bands are expected for the alcohols?

4. Draw the possible aldehydes and ketones with the formula C_4H_8O. What principle absorption bands are expected for the aldehydes and ketones?

Introduction to Organic Analysis: Infrared Spectroscopy

Name:	Lab Instructor:
Date:	Lab Section:

RESULTS and POST-LABORATORY QUESTIONS

Part A. Functional Groups

Attach a copy of the Lewis dot structures for molecules A – G that you drew in your laboratory notebook. On your drawings, identify the hybridization of each carbon atom.

Using the table below, name the functional group(s) present in each molecule (A-G) and identify the expected range(s) of IR absorption frequencies.

Compound	Functional groups	Absorption frequency
A		
C		
E		
G		

Compound	Functional groups	Absorption frequency
B		
D		
F		

Part B. Detective Work

Scenario 1:

Based only on Analysis #1 and #2, what is the molecular formula of the unknown compound? Show all your work.

Which chemicals A – G (in Part A) have the required molecular formula?

Based on your answer to the previous question and the IR from Analysis #3, what is the correct structure for the compound? Explain.

OVER →

RESULTS and POST-LABORATORY QUESTIONS continued...

Scenario 2:
In which IR region(s) do nitriles exhibit characteristic absorption bands? Explain.

Scenario 3:
Attach a copy of the Lewis dot structures for molecules H – M that you drew in your laboratory notebook.

Identify each of the six unknowns by matching the letter (H – M) and the corresponding formula to the following spectra. Briefly justify how you made each assignment.
Spectrum 1:

Spectrum 2:

Spectrum 3:

Spectrum 4:

Spectrum 5:

Spectrum 6:

Colligative Properties: Analysis of Freezing Point Depression

Kristen Spotz

OBJECTIVES

- Verify experimentally that colligative properties depend on the number of solute particles and not on the solute's identity.
- Perform a detailed analysis of the freezing point depression of cyclohexane.

INTRODUCTION

You wake up one chilly January morning. The temperature is −5.0°C outside and you wonder whether your car will start. You walk outside and find your car surrounded by snow. To your surprise your car does start thanks to the antifreeze, ethylene glycol ($C_2H_6O_2$), you added to your car's radiator (Figure 1).

Figure 1. Adding antifreeze to a car's radiator is a necessity in cold climates

Similarly, arctic fish require a form of biological antifreeze, typically glycerol ($C_3H_8O_3$), to help prevent their blood from freezing (Figure 2).

Who would have thought that an automobile radiator and an arctic fish would have so much in common?

Figure 2. A cold-loving arctic fish.

Background

When a non-volatile solute is added to a solvent, the resulting solution has different physical properties than the pure solvent. Going back to our antifreeze example, the pure solvent, water, has a freezing point of 0°C. When we add a non-volatile solute such as ethylene glycol, CH_2OHCH_2OH, the freezing point of the solution is lowered. Surprisingly, the amount the freezing point lowers depends only on the quantity of solute particles added and not on the actual identity of the solute added. There are four such properties, known as colligative properties, that are dependent on the amount of solute: vapor pressure lowering, boiling point elevation, freezing point depression, and osmotic pressure. The basis for the colligative properties lies in the inability of the solvent particles to join the solvent in moving between phases or through a semipermeable membrane, in the case of osmotic pressure.

Since the focus is on the number of solute particles added, we must account for the fact that in cases where the solutes are electrolytes, the formula of the solute must be taken into account. For example adding 1.0 mole of NaCl to a liter of water has the effect of actually adding 2.0 moles of particles (1.0 mole of Na^+ ions and 1.0 mole of Cl^- ions).

Vapor Pressure Lowering

The vapor pressure of a solution containing a non-volatile solute is always less than the vapor pressure of the pure solvent. This concept can be explained in terms of entropy, which is the measure of the system's disorder. Processes in nature tend to occur in the direction of increasing disorder. The molecules in the liquid phase are closer together and more ordered than the molecules in the gas phase. Therefore, the entropy of a pure liquid increases as the molecules vaporize. This process continues until the system reaches a dynamic equilibrium at which point the molecules are vaporizing and condensing at the same rate. At this point the vapor pressure is accessed (Figure 3a). However, when a nonvolatile solute is dissolved in a solvent, the resulting solution is more disordered than the pure liquid with which you began. The solvent in the solution has less of a tendency to vaporize and equilibrium is established at a lower vapor pressure (Figure 3b).

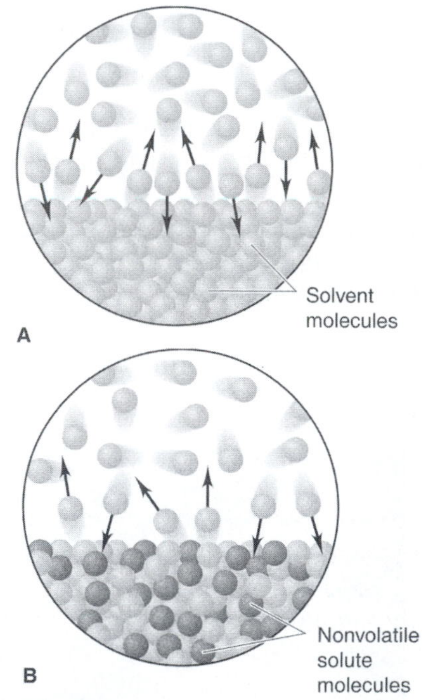

Figures 3a,b. The effect of solute particles on the vapor pressure of a liquid.

We can also consider the decrease in the vapor pressure of a solution in more simplistic terms. The dissolved nonvolatile solute decreases the number of solvent molecules on the solution's surface that are capable of vaporizing. In a sense, the solute acts to "block" the solvent molecules in the liquid phase from entering the vapor phase, resulting in a decreased vapor pressure.

Boiling Point Elevation

Under normal conditions, pure water boils at 100.00 °C. However, if 18.0 grams of glucose is dissolved in 150 mL of water, the resulting solution boils at 100.34 °C. This increase in boiling point can be explained by examining the phase diagram for water (Figure 4). The boiling point of a liquid is the temperature where the vapor pressure of the liquid equals the atmospheric pressure. At this point, the liquid and vapor phase molecules are in equilibrium. Start by locating the temperature corresponding to the boiling point of water in

Figure 4. Notice that at this temperature, the vapor pressure of the water is 1 atm. As long as the atmospheric pressure is also 1 atm, the water will boil. You should notice that at the same temperature, the vapor pressure of the solution (dotted line) is less than 1 atm. Therefore, to boil the solution, temperature must be increased (by ΔT_b) in order to raise the vapor pressure of the solution until it is also equal to the atmospheric pressure.

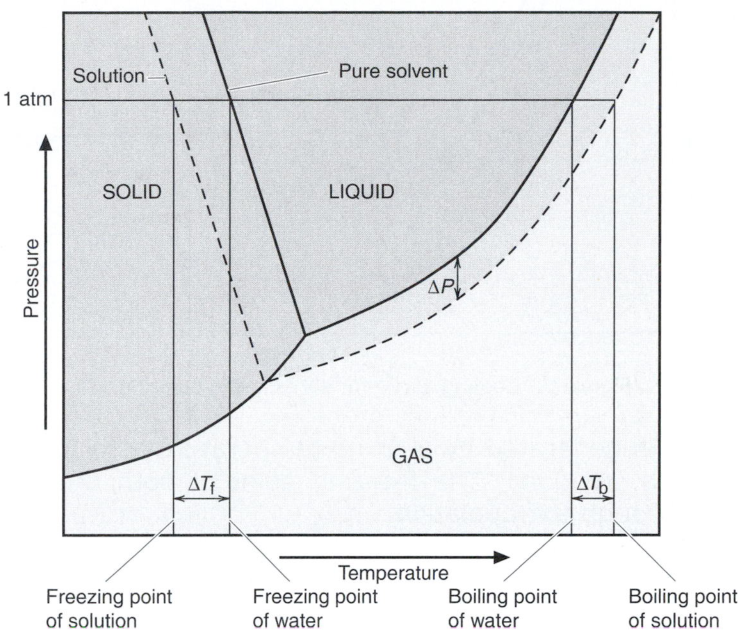

Figure 4. The phase diagram of pure water overlaid with the phase diagram of an aqueous solution.

The magnitude of the boiling point elevation can be calculated using Equation 1:

$$\Delta T_b = K_b m \qquad \text{Equation 1}$$

In Equation 1, ΔT_b is the difference between the boiling point of the pure solvent and the boiling point of the solution (°C), K_b is the molal boiling-point-elevation constant (°C/m), which is specific for each solvent, and m is the <u>molality</u> of the solution. Scientists can utilize Equation 1 to identify the molar mass of an unknown solute or to predict how much solute to add to obtain a required temperature change.

Freezing Point Depression

Freezing point depression is quantified by an equation similar in form to Equation 1. The actual freezing point is difficult to determine, however, by direct visual observation because of a phenomenon known as supercooling. With supercooling, a liquid being slowly cooled can remain liquid below its freezing point because the molecules are unable to orient themselves into the crystalline array characteristic of a solid. Constant stirring is utilized to reduce the effect of supercooling and to prevent an incorrect freezing point reading. In addition, a temperature-time graph or cooling curve (Figure 5) will be used to more accurately determine the freezing points of the solvent and solution.

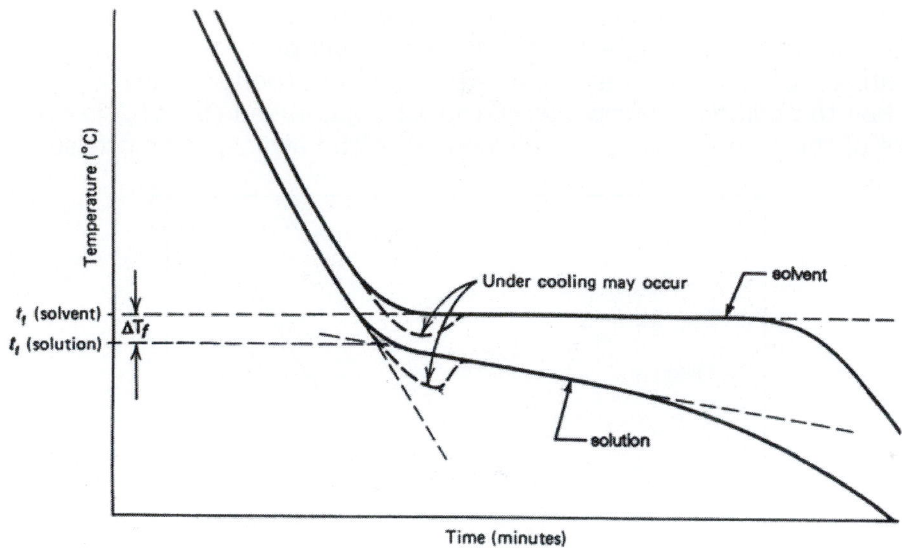

Figure 5. Cooling curve for solvent and solution.

The above cooling curve is generated by plotting temperature as a function of time for both the pure solvent and the solution. The freezing point of both the pure solvent and the solution is determined through extrapolation.

OVERVIEW

The nature of colligative properties will be explored by observing what happens to the freezing point of water after adding several different solutes. Your in depth study of the freezing point depression of cyclohexane will begin by constructing a cooling curve for pure cyclohexane. Half the class will then construct a cooling curve for a solution of naphthalene in cyclohexane while the other half of the class constructs a cooling curve for a solution of an unknown solute in cyclohexane. Information drawn from these curves will allow for the calculation of the molal freezing-point-depression constant for cyclohexane as well as the molar mass of the unknown solute.

PROCEDURE

Part A: Confirming the Nature of Colligative Properties

Chemicals Used	Materials Used
Urea (1.8 g)	Styrofoam cups (3)
Glucose (5.4 g)	Analytical balance
Sodium chloride (0.90 g)	Spatula
Ice	Thermometer

All students, working in pairs

Half-fill a Styrofoam cup with ice. Record the initial temperature of the ice. Add 1.8 g of urea to the cup. Stir until the urea is dissolved. Record the final, lowest temperature of the mixture. Repeat with 5.4 g of glucose and again with 0.9 g of sodium chloride. Dispose of the urea solution according to your instructor's directions.

Part B: Determining the Cooling Curve for Pure Cyclohexane

Chemicals Used	Materials Used
Cyclohexane NaCl Ice	200-mm test tube 400-mL beaker 600-mL beaker 20-mL graduated cylinder Thermometer Ring stand Test tube clamp Stopwatch

All students, working in pairs

1. Place a clean, dry 200-mm test tube in a 400-mL beaker. Record the mass. Using the density of cyclohexane (0.7739 g/mL), estimate the volume needed for about 10 grams of cyclohexane. Pour this volume into the test tube. Reweigh the test tube, the beaker and the cyclohexane. Wrap a paper towel around the 400-mL beaker and place inside the 600-mL beaker. Fill the 400-mL beaker completely full with ice. Add approximately 2 grams of NaCl to the ice.

2. Place the 600-mL beaker containing the ice bath and the test tube on a ring stand. Attach a test tube clamp to the ring stand to hold the test tube. Lower the thermometer completely into the test tube filled with cyclohexane.

3. While **constantly stirring** the cyclohexane with a thermometer, record the temperature (±0.2°C) at 20 second intervals. Once the cyclohexane solidifies, the temperature remains virtually constant. Continue collecting data at 20 second intervals until the temperature begins to drop again (typically after a total of 10 minutes).

4. Allow the cyclohexane to melt before reusing in Part C or D.

Part C: Determining the K_f for Cyclohexane

Chemicals Used	Materials Used
Naphthalene ($C_{10}H_8$) NaCl Ice	Cooling curve apparatus (from Part B) Spatula Analytical balance Stopwatch

Half the class, working in pairs

Measure approximately 0.30 grams of naphthalene (record the actual mass used) and transfer into the test tube filled with cyclohexane that you used in Part B. Make sure the naphthalene is completely dissolved before proceeding to the next section. Add ice and salt to the 400-mL beaker as needed.

Determine the freezing point of the solution in the same way you did for the pure solvent. Again, be sure to collect data for at least 10 minutes. Dispose of the cyclohexane solution according to your instructor's directions.

Part D: Determining the Molar Mass of an Unknown Solute

Chemicals Used	Materials Used
Unknown solute NaCl Ice	Cooling curve apparatus (from Part B) Spatula Analytical balance Stopwatch

Half the class, working in pairs
Measure approximately 0.30 grams of unknown (record the actual mass used) and transfer into the test tube filled with cyclohexane that you used in Part B. Make sure the unknown is completely dissolved before proceeding to the next section. Add ice and salt to the 400-mL beaker as needed.

Determine the freezing point of the solution in the same way you did for the pure solvent. Again, be sure to collect data for at least 10 minutes. Dispose of the cyclohexane solution according to your instructor's directions.

Part E: Group work

All students, working in pairs
Students who did Part C: graph your data. Determine the freezing point. From the data in Part B and Part C, determine the freezing point depression. Write your value of K_f on the board with the class data. Before you leave, copy all of the class data.

Students who did Part D: graph your data. Determine the freezing point. From the data in Part B and Part D, determine the freezing point depression. Using the class average for K_f (from Part C), calculate the molar mass of the unknown. Before you leave, write your value for the molar mass on the board and copy all of the class data.

Colligative Properties: Analysis of Freezing Point Depression

Name:	Lab Instructor:
Date:	Lab Section:

PRE-LABORATORY EXCERCISES

1. Define the underlined terms in the BACKGROUND section.

2. Is the following statement TRUE or FALSE: "Dissolving a nonvolatile solute in a solvent has the effect of extending the temperature range over which the solvent remains in the liquid phase." Explain.

3. Write the equation for the freezing point depression (similar to Equation 1).

4. Describe a practical application of the freezing point depression (other than the ones already given in the INTRODUCTION and BACKGROUND).

OVER →

PRE-LABORATORY EXCERCISES continued...

5. Assuming that each system behaves ideally, which solution (0.25 m KCl or 0.25 m $C_3H_8O_3$) would have the lower freezing point? Explain.

6. What is the boiling point of the water in your radiator if 2.00 kg of antifreeze (ethylene glycol, $C_2H_6O_2$) is added to 9.00×10^3 grams of water?

Colligative Properties: Analysis of Freezing Point Depression

Name:	Lab Instructor:
Date:	Lab Section:

RESULTS and POST-LABORATORY QUESTIONS

Part A: Confirming the Nature of Colligative Properties

Solute	Formula	Grams used	Moles of particles	Initial Temp.	Final Temp.
Urea					
Glucose					
Sodium Chloride					

What conclusions can you draw by comparing the freezing points of the solutions of urea and glucose? Explain.

What conclusions can you draw by comparing the freezing points of the solutions of urea and sodium chloride? Explain.

Part B: Determining the Cooling Curve for Pure Cyclohexane

Mass of cyclohexane used	
Freezing point of cyclohexane (from curve)	

Attach a copy of your cooling curve

Part C and D

Which part did you and your laboratory partner perform? (Circle one) Part C or Part D

	Part C	Part D
Freezing point of solution (from curve)		
Freezing pt. depression, ΔT_f		
Mass of solute added		
Moles of solute added		
Mass of solvent, kg		
Molality of solution		
Your value of K_f for cyclohexane		
Class average for K_f for cyclohexane		
Your value of molar mass of unknown		
Class average of molar mass of unknown		

Attach a copy of your cooling curve

OVER →

RESULTS and POST-LABORATORY QUESTIONS continued...

1. Which of the following substances would be most efficient per unit mass at melting snow from sidewalks and roads: glucose, sodium chloride, ethylene glycol, or calcium nitrate. Your explanation should include possible environmental impact.

2. Pure benzene has a normal freezing point of 5.50°C. A solution containing 11.4 grams of a molecular substance dissolved in 150.0 grams of benzene (K_f = 5.12°C/m) has a freezing point of 1.20°C. What is the molar mass of the solute?

3. If you had incorrectly read the freezing point of the solution in POST-LABORATORY QUESTION #2 to be 0.30°C lower than the true freezing point of the solution, would the calculated molar mass of the solute by too high or too low? Explain.

Introduction to Kinetics: Factors That Affect the Rate of Reaction

Chad Eller

OBJECTIVES

- Be able to list and rationalize the factors that affect the rates of a reaction.
- Explain various scenarios using the factors that affect reaction kinetics.

INTRODUCTION

Throughout nature, chemical reactions occur at different rates. Some reactions such as the rusting of iron are relatively slow while others such as the combustion of gasoline occur very quickly. Scientists, however, have figured out ways to make various reactions run faster or slower. Becoming familiar with the factors that affect the rate of a reaction gives us insight into how reactions work. The field of chemistry that is concerned with the rate at which reactions occur is called chemical kinetics.

A rough analogy can be made to the speed with which a computer completes a specific task. All computers are not created equal. If you try to run the newest 3-D game in high resolution on an old machine you'll be lucky to get it to work at all. Each component of a computer has a definite and predictable effect on its performance. Too little RAM, slow bus speed, fragmented hard drive, inefficient operating system or application, multitasking, network congestion – they all work to slow down our computing experience. But for each problem there is a solution. It just takes a little knowledge (theory) and a few tries at improvement (experiments).

Chemists and computer engineers are not the only people concerned with the rates of processes. Consider these examples:

Career	Application
Biologist	Preservation or decomposition of specimens
Chemical Engineer	Speed of production effecting cost
Civil Engineer	Concrete and asphalt curing
Doctor	Medication or poison effecting the body
Museum Curator	Dating, restoration, preservation of artifacts
Restaurant Owner	Food spoilage and safety

BACKGROUND

In order to understand each factor that affects the rate of a chemical reaction we can use the simple model of atoms as very small spheres in constant motion. Molecules are groups of these spheres that are bonded together and are constantly bouncing off each other. Picture just a few molecules at a time and consider what happens to them in different circumstances. Think of this model as we study the effects of concentration, surface area (for solids), temperature, and catalysts.

Effect of Changing the Concentration of Reactants

Chemical reactions involve breaking chemical bonds, rearranging the reactant atoms, and making new chemical bonds. In order for this to occur, molecules must collide with each other. If there are only a few molecules of each reactant in a given volume, the number of collisions between them will be relatively low. By increasing the concentration of the reactants, we increase the number of reactant molecules in the same amount of space. This means there are more opportunities for a collision to occur.

Fishing can be used as an analogy for the effect of concentration. If you are fishing in a well-stocked pond with thousands of fish (a high concentration) you have a better chance of catching a fish than if the same pond only had 2 or 3 fish (a lower concentration).

Effect of Changing the Surface Area

In the case where one of the reactants is a solid, the majority of the atoms are trapped beneath the surface. Only the atoms on the surface are available to collide with the other reactant. When a sample cube is cut into smaller pieces (Figure 1), the amount of surface area increases, even though the volume does not change. Grinding a solid into a powder vastly increases the surface area, making a larger portion of the atoms available to collide with the other reactant. In your daily experience, you may have seen that fine salt crystals dissolve in water faster than course rock salt.

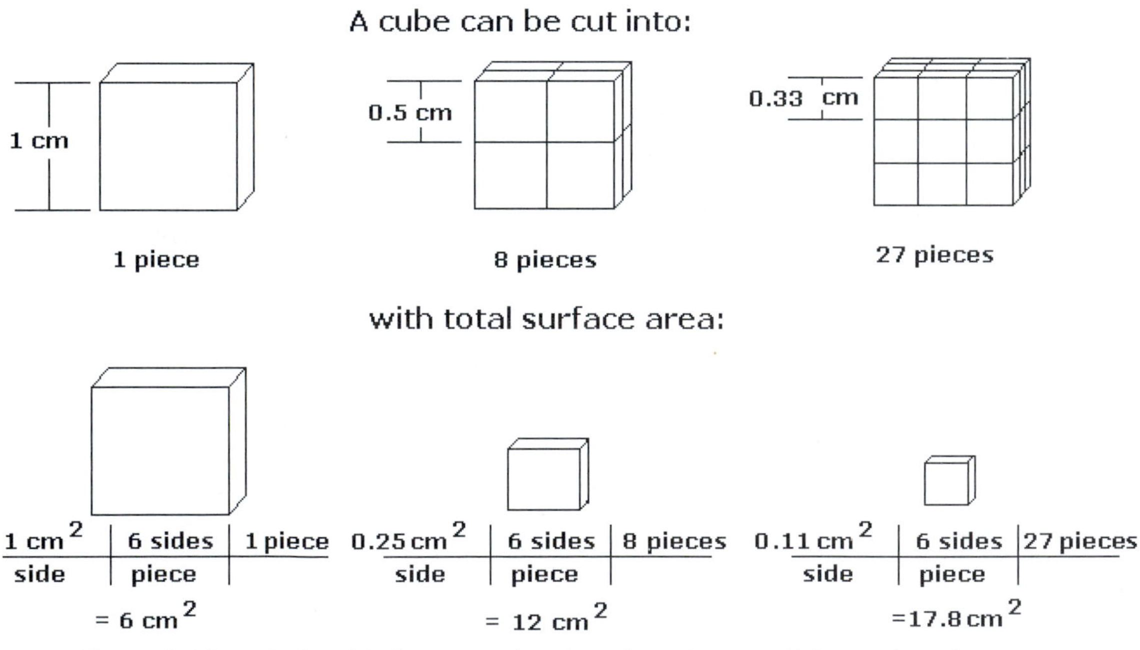

Figure 1. The relationship between the size of reactant particles and surface area

Effect of Changing the Temperature

The average molecular kinetic energy of a sample is constant at a given temperature. However, the random nature of molecular motion means that some molecules will be moving faster than others. At any given temperature a few molecules have enough energy to react. This minimum required energy is called the activation energy, E_a. As the temperature of the system is increased, the kinetic energy available during collisions goes up and the proportion of collisions exceeding E_a increases (Figure 2). This allows the reaction to take place faster at a higher temperature.

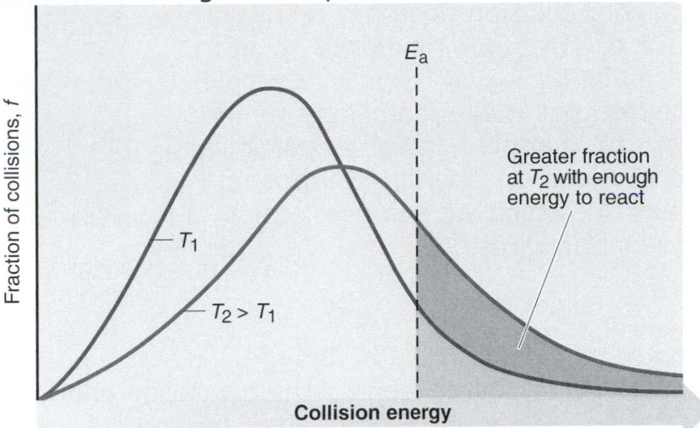

Figure 2. The relationship between temperature and the fraction of collisions with enough energy to react

Think of rolling a ball up an inclined driveway into a garage. If you roll the ball slowly, it comes right back to you. When you roll it fast enough, however, the ball makes it into the garage. The amount of energy needed to get the ball into the garage is analogous to the activation energy of a reaction. Only when a molecule can acquire at least that much energy does a reaction take place.

Just as some driveways are steeper than others, chemical reactions differ in the amount of energy needed to make them occur. Why don't all reactions have the same E_a? Sometimes the bond to be broken is very strong. In other reactions there is an unstable intermediate molecule that requires a lot of energy to make (Figure 3).

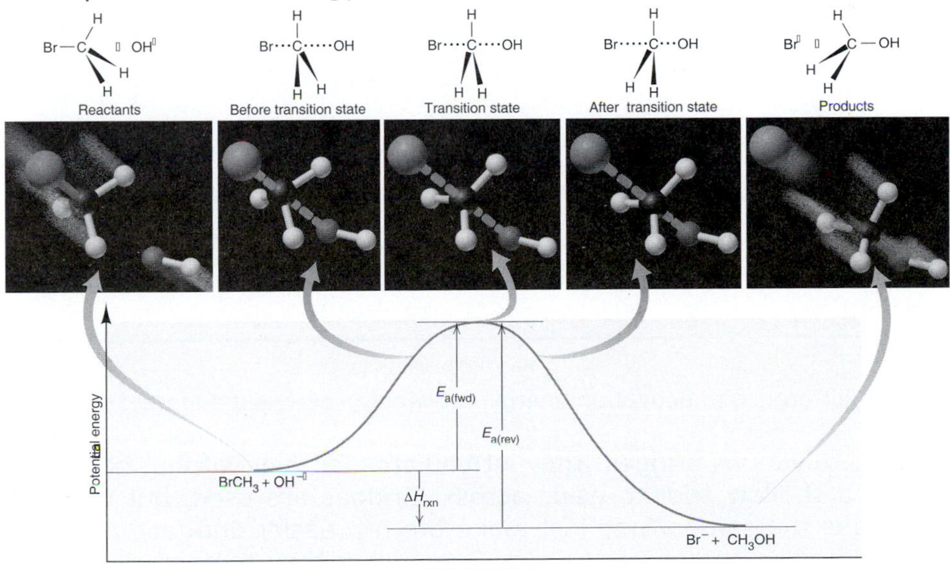

Figure 3. Activation energy and transition state for the reaction between CH_3Br and OH^-

Orientation of the Collisions

As every baseball player knows, hitting the ball does not guarantee a home run. Sometimes the ball hits the top of the bat and pops straight up, other times it hits the bottom of the bat, and the batter grounds out. Only when the swing is perfectly aligned with the ball can you hit a home run.

Likewise, in chemistry, every collision does not result in a chemical reaction (Figure 4). In order for bonds to form, atomic orbitals must overlap just right. Complex molecules can have shapes that make it unlikely for this overlapping of orbitals to happen in any particular collision. The likelihood of a correct spatial relationship is expressed in the constant 'A', which we will use later in a mathematical model of reaction rates.

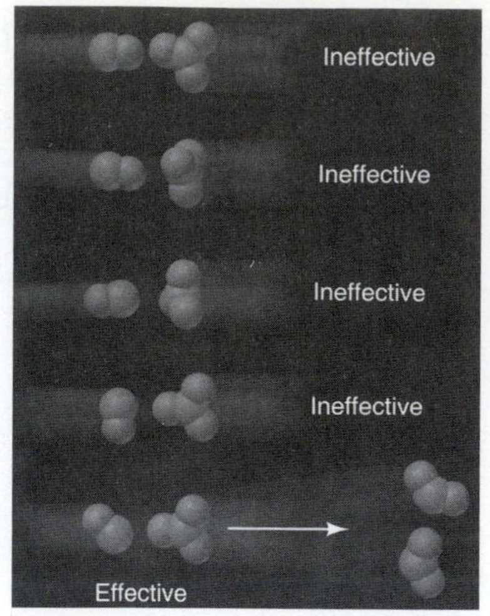

Figure 4. The importance of correct collision orientation on the success of a chemical reaction.

Effect of Adding a Catalyst

Although scientists are not able to directly control the activation energy or the orientation of a collision for a reaction, the use of catalysts often allow for the manipulation of these factors. A catalyst is a material that does not permanently change or get used up in a reaction, but helps the reaction run faster. The catalyst lets a reaction form the same product it normally would, but by following a different, less energy intensive route (Figure 5).

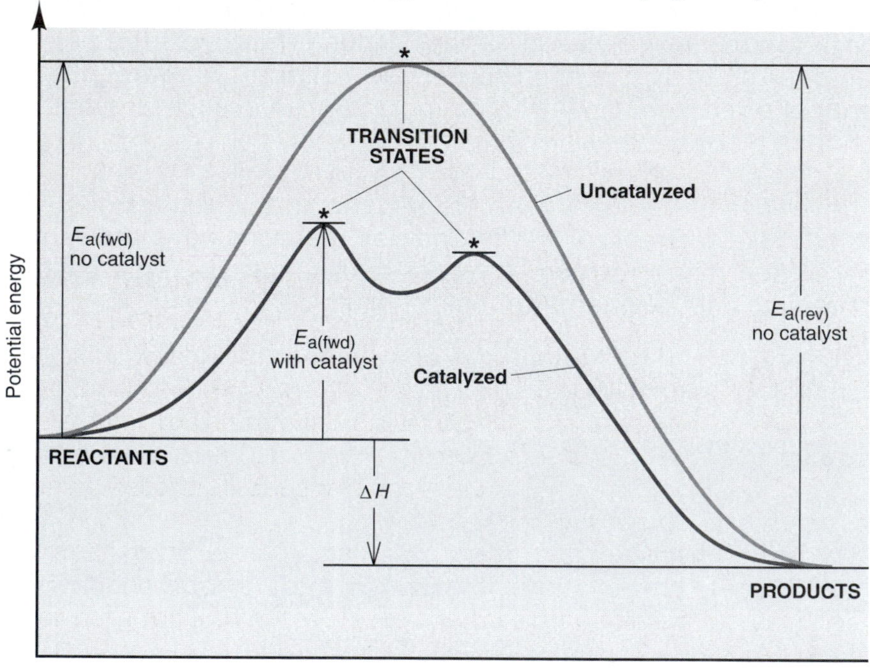

Figure 5. The difference in activation energy between a catalyzed and uncatalyzed reaction.

You can think of catalysts as bridges: they let you cross a river without having to walk miles and miles to find a shallow spot to wade across. Bridges are used, but not destroyed, and where you end up is the same. They just make the trip easier and faster. Catalysts work in many different and often complicated ways. Sometimes they temporarily donate or absorb electrons, hydrogen ions or hydroxide ions in order to provide the alternative, lower energy reaction mechanism.

OVERVIEW

Each factor that affects the rate of reaction will be demonstrated with a chemical reaction or model. Pay attention to which factor is affecting the rate in each case. A key to understanding kinetics is to consider how the changes we make will determine what the atoms are experiencing thereby leading to a change in the rate of reaction.

PROCEDURE

Part A. Effect of Changing the Concentration of Reactants

Chemicals Used	Materials Used:
HCl, 1M and 6M in dropper bottles Chalk	Watch glass (2)

CAUTION: 6M HCl is caustic. If any is spilled on your skin, immediately rinse with running water and inform your laboratory instructor. As always, you should wear your goggles at all times when working in the laboratory. Dispose of waste according to instructor's directions.

Place a small piece of chalk in each of the two watch glasses. Add 10 drops of 1M HCl to one sample of chalk and 6M HCl on the other. Record your observations.

Part B. Effect of Changing the Surface Area

Chemicals Used	Materials Used:
Coffee creamer $CuSO_4$, 0.2 M Steel wool, non-detergent	Candle (tea light or votive candle) and lighter Spatula Disposable pipet 50-ml beaker (3) Hot plate and thermometer Glass stirring rod (2)

CAUTION: Clear flammable materials from lab bench and surrounding areas.

Case I: Flammability of coffee creamer

After lighting the candle, use a spatula to hold a small amount of coffee creamer in the flame for 5 seconds. Record your observations. Next, draw a small amount of coffee creamer into a disposable pipet. While standing at arms length from the candle, aim a burst of coffee creamer at the flame. Again, record your observations. Clean your lab bench according to your instructor's directions.

Case II: Reaction between $CuSO_4$ and steel wool

Add 25 mL of 0.2 M $CuSO_4$ to each of two 50-mL beakers and heat both to 80°C. While the solution is heating, prepare two pieces of steel wool. Each piece should be about 0.2 grams. The first piece should be stretched out and the second piece should be rolled between your fingers until it becomes a tight ball. Simultaneously drop the pieces of steel wool into the two beakers and stir. Record any color changes and how long it took before the changes occurred. Dispose of waste according to your instructor's directions.

Part C. Effect of Changing the Temperature

Chemicals Used	Materials Used:
Food coloring (dark) $CuSO_4$ (0.2 M) Ice Zinc metal (10 mesh, granular)	400-mL beaker (2) Hot plate and thermometer Tongs or heat resistant gloves Spatula 50-mL beaker (3) 100-mL beaker Clay, small lump

Case I: Dispersal of dye in hot and cold water

Fill two 400-mL beakers each with 250 mL water. Heat one to 80°C. Without stirring, add one drop of food coloring to each container by touching the surface with the dropper. Record the time required for the color to disperse.

Case II: Reaction between $CuSO_4$ and zinc

Add 25 mL of 0.2 M $CuSO_4$ to each of two 50-mL beakers. Heat one solution to 80°C, while cooling the other in an ice bath. Add a few pieces of granular zinc to the cold container. Record any color changes and how long it took for the changes to occur. Repeat by adding a few pieces of zinc to the hot solution. Again, record your results. Dispose of waste according to your instructor's directions.

Case III: Modeling Activation Energy

Roll a piece of modeling clay into a ball no bigger than a ping-pong ball. Drop the clay on a clean area of floor from a height of 1 foot. Gently push the clay sideways to see if it rolls. Note any shape change upon impact. Drop the clay several more times each time about one foot higher. Roughly how high did you need to drop the clay for it to stick firmly to the floor?

Part D. Modeling the Significance of the Orientation of Collisions

Materials Used:
Styrofoam-Velcro balls (4, 2 each with one piece of Velcro and 2 each with 6 pieces of Velcro) Box or deep tray

Place two of the Styrofoam balls with 6 pieces of Velcro on them in a box or deep tray. Gently shake the container until the two balls stick. Repeat with the 2 balls that have only one Velcro square. Describe how readily the balls stick together in each case.

Part E. Effect of Adding a Catalyst

Chemicals Used	Materials Used:
Hydrogen peroxide (H_2O_2), 6% solution KI Ice 0.5% corn starch solution Iodine solution (saturated) in dropper bottles	100-mL beaker Hot plate and thermometer Tongs or heat resistant gloves 600-mL beaker Spatula 50-mL beaker (3) Glass stirring rod (2)

CAUTION: Avoid contact with hydrogen peroxide. If any is spilled on your skin, immediately rinse with running water and inform your laboratory instructor. As always, wear your goggles at all times when working in the laboratory. Dispose of peroxide waste according to instructor's directions. In general, peroxide waste is NOT compatible with other chemical waste.

Case I: Decomposing H_2O_2

In a 100-mL beaker, heat 20 ml of 6% H_2O_2 to 80°C. Using tongs or heat resistant gloves, remove the beaker from the hot plate and place the whole beaker inside a large 600-mL beaker. Using a spatula, add roughly 0.05 g of KI (a pinky nail sized portion) to the H_2O_2. Record your observations. Do not view from above, as minor splashing may occur. After 30 - 45 seconds, add 10 mL of ice to the solution. Again, record your observations.

Case II: Hydrolysis of Starch

Using 50-mL beakers, warm two, 10-mL samples of 0.5 % corn starch solution to 40°C. Add 10 drops of iodine to each sample and record your observations. Collect 2 mL of saliva from yourself and lab partner. Thoroughly mix the saliva into one of the corn starch/iodine solutions. Record the time required for the solution to become light lavender or clear. Add a few more drops of iodine. Note the color of drops as they enter solution, and the color after mixing. Dispose of waste according to your instructor's directions.

Introduction to Kinetics: Factors That Affect the Rate of Reaction

Name:	**Lab Instructor:**
Date:	**Lab Section:**

PRE-LABORATORY EXERCISES

1. Name four factors that are under the control of a scientist when he or she wants to increase the rate of a reaction.

2. As scientists, we often talk about wanting to speed reactions up. Can you think, however, of any reactions that you might rather slow down than speed up? Can you imagine any way to use the factors from the previous question to help you control the specific reaction you want to slow down?

3. You have an important project due tomorrow and you are trying to use every free second to put on the finishing touches. Whenever possible, you also rush through your typical daily tasks and responsibilities. Which of the following has associated with it a time constraint that is under your control and can be adjusted safely: showering, driving to school, math class, eating lunch, Doctor's appointment and grocery shopping. Briefly explain your answers.

4. A living cell must accomplish many complicated chemical tasks. Our bodies contain countless enzymes that are used to speed up otherwise slow reactions. Knowing that enzymes are a type of catalyst, which of the following factors is effected by the presence of an enzyme: the energy of the reactants, the energy of the products, the energy of the transition state, the likelihood of a collision with the correct orientation, or the temperature of the reaction. Briefly explain your answer(s).

Introduction to Kinetics: Factors That Affect the Rate of Reaction

Name:	**Lab Instructor:**
Date:	**Lab Section:**

RESULTS and POST-LABORATORY QUESTIONS

Part A. Effect of Changing the Concentration of Reactants

In what ways was the reaction between the chalk and the 1 M HCl similar to the reaction between the chalk and the 6 M HCl? In what ways were the reactions different?

Part B. Effect of Changing the Surface Area

Case I: Flammability of coffee creamer

What was the effect of spraying the coffee creamer at the flame rather than holding a spatula full in the flame? Explain your observations in terms of surface area.

Case II: Reaction between $CuSO_4$ and steel wool

Describe the appearance of the solution before addition of the steel wool.

Describe the appearance of the solution and the steel wool after the reaction.

In which case did the reaction occur first? Explain why.

Part C. Effect of Changing the Temperature

Case I: Dispersal of dye in hot and cold water

Time required to disperse dye in room temperature water:

Time required to disperse dye in hot water:

Explain your observations in terms of the kinetic energy of the water molecules.

OVER →

RESULTS and POST-LABORATORY QUESTIONS continued...

Case II: Reaction between $CuSO_4$ and zinc

	Cold Solution	Hot Solution
Time for 1st color change in the...		
Time for 2nd color change in the...		

What effect does increasing the temperature have on the rate of reaction? Explain why.

Case III: Modeling Activation Energy

Why doesn't the clay always stick to the floor? Explain how this activity serves as an analogy for activation energy.

Part D. Modeling the Significance of the Orientation of Collisions

In which case (the ball with 1 Velcro or 6 Velcro pieces) was the required orientation for successful collision more restrictive? Explain.

Did this agree with your observations of how long it took for the balls to stick? Explain.

Part E. Effect of Adding a Catalyst

Case I: Decomposing H_2O_2

Observations after addition of KI:

Observations after addition of ice:

In Case I, which material is the catalyst?

Does temperature affect the usefulness of a catalyst?

RESULTS and POST-LABORATORY QUESTIONS continued...

Case II: Hydrolysis of Starch

Description of corn starch solution:

Observation after addition of Iodine:

Observation after addition of saliva (compare with control):

Observation after addition of more Iodine:

1. Marble, like chalk is composed of $CaCO_3$. Explain why monitoring the acidity of rainfall would be important with regards to conserving historically and artistically important outdoor statues.

2. A chemical engineer is trying to increase output of a chemical plant. She is considering using an expensive catalyst or increasing the temperature of the large reaction vessel by 20°C to accomplish the same task. Which route will be least expensive in the short term? Long term?

3. Is the blue chemical in the $CuSO_4$ solution a catalyst? How do you know?

4. As a reaction progresses and the reactants are consumed, will this tend to increase or decrease the rate of reaction? Explain.

Determining the Rate Law: A Kinetics Study of the Iodination of Acetone

Kristen Spotz

OBJECTIVES

- Gain a quantitative understanding of kinetics.
- Determine the rate of a reaction, the order of the reaction with respect to the reactants and the value of the rate constant.
- Predict reaction times using an experimentally determined rate law.

INTRODUCTION

One factor influencing the rate of a reaction is the concentration of the reactants. Typically, as the concentration of reactants increases so does the rate of a reaction. The actual relationship, however, can be quite complicated. Sometimes, doubling the concentration will result in a doubling of the rate. For other chemical reactions, however, doubling the concentration of a reactant might have no effect on the rate or it might result in a four-fold increase in the rate. It can be extremely useful for a scientist to have an understanding of the relationship between the concentration of the reactants and the rate of the reaction. For example, with a quantitative knowledge of reaction rates, scientists are able to gain insight into reaction mechanisms and even predict the time frame of a reaction. Detailed studies by 1995 Nobel laureates F. Sherwood Rowland and Mario Molina of the rates of hundreds of chemical reactions provided insight into the role of chlorfluorocarbons (CFCs) in the depletion of the ozone layer. So, how do scientists determine the relationship between concentration and reaction rate and how are the results expressed in a useful form?

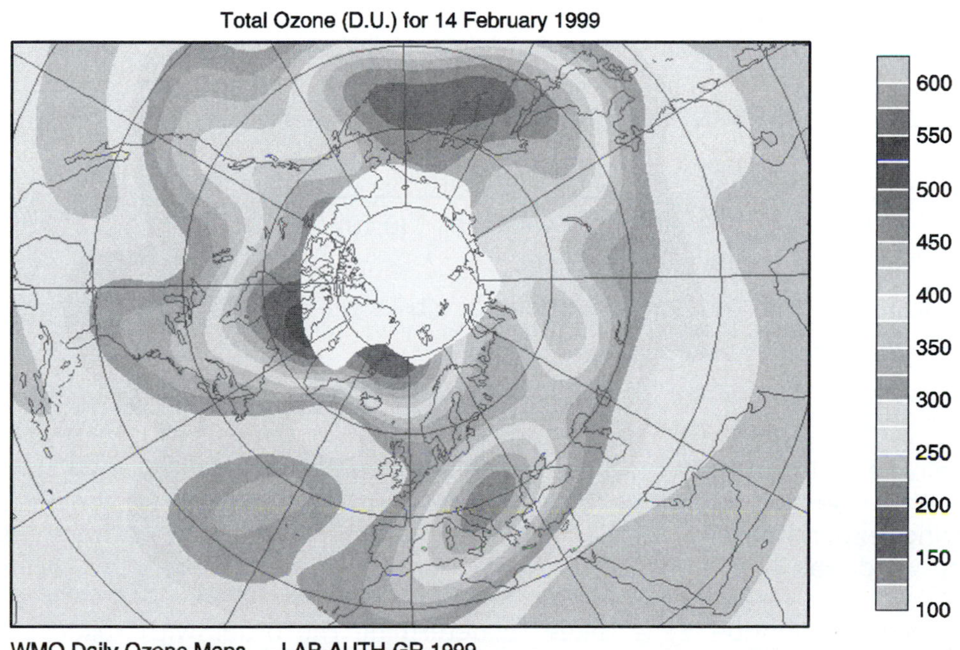

Figure 1. A map of the ozone hole over the Northern Hemisphere, February 14, 1999. World Meteorological Organization: www.wmo.ch/web/arep/nhoz/html.

BACKGROUND

The Rate Law

All the information needed to predict the rate of a given reaction is contained in the rate law, or the rate equation. Given the generic reaction in Equation 1, the general form of the rate law would be given in Equation 2:

$$aA + bB \rightarrow cC + dD \qquad \text{Equation 1}$$

$$\text{rate} = k[A]^m[B]^n \qquad \text{Equation 2}$$

The rate constant, k, is specific for each reaction and is temperature dependent. The units for k are dependent on the overall order of the reaction. The rate law also includes the concentrations of the reactants raised to the reaction orders, m and n. The values for m and n must be determined experimentally (as you will do in today's experiment) and cannot be derived from the balanced chemical equation.

As an example, we will examine the reaction (Equation 3) and the experimental data (Table 1) for the reaction of nitrogen monoxide and hydrogen gas to produce nitrogen gas and steam.

$$2NO\ (g) + 2H_2\ (g) \rightarrow N_2\ (g) + 2H_2O\ (g) \qquad \text{Equation 3}$$

Table 1

Exp. #	$[NO]_{initial}$, M	$[H_2]_{initial}$, M	Initial rate, M/s
1	0.10	0.10	1.23×10^{-3}
2	0.10	0.20	2.46×10^{-3}
3	0.20	0.10	4.92×10^{-3}

The first step in determining the rate law is to follow the example in Equation 2 and write the general form of the rate law for the reaction (Equation 4):

$$\text{Rate} = k[NO]^m[H_2]^n \qquad \text{Equation 4}$$

The next step is to find the order for each reactant. In order to find the order with respect to hydrogen gas, Experiment #1 and #2 are compared:

$$\frac{\text{Rate 2}}{\text{Rate 1}} = \frac{2.46 \times 10^{-3}}{1.23 \times 10^{-3}} = \frac{k[0.10]^m[0.20]^n}{k[0.10]^m[0.10]^n} \qquad \text{Equation 5}$$

The rate constants, k, and the concentrations of nitrite will cancel (along with the unknown m) leaving Equation 6:

$$2 = 2^n \qquad \text{Equation 6}$$

Mathematically, n = 1. In terms of kinetics, this is interpreted to mean that the reaction is first order with respect to hydrogen gas. This result makes sense if we look back at Table 1. Comparing experiments 1 and 2, if the concentration of nitrogen monoxide is held constant and we double the concentration hydrogen, the rate of the reaction doubles.

The same process (Equation 5) is taken to determine the order with respect to nitrogen monoxide. From setting up the ratio of experiments 1 and 3, the order with respect to nitrogen monoxide is determined to be second order. This means that if the concentration of hydrogen is held constant while doubling the concentration of nitrogen monoxide, the rate of the overall reaction quadruples. The overall order of the reaction (n+m) is 3.

Now that we know the order of the reaction, the next step is to determine the value of the rate constant, k. The units for k are dependent on the overall order of the reaction. Data from any experiment given in the table may be used to determine the rate constant. For example, using the data from experiment #1 results in Equation 7.

$$1.23 \times 10^{-3}\ M/s = k[0.10\ M][0.10\ M]^2 \qquad \text{Equation 7}$$

Solving for the rate constant, the value is equal to 1.23 $M^{-2}s^{-1}$. The final rate law includes the value for both the rate constant and the orders of the reaction.

$$\text{Rate} = 1.23\ M^{-2}s^{-1}\ [H_2][NO]^2 \qquad \text{Equation 8}$$

Using Equation 8 above, any initial concentration of hydrogen gas and nitrogen monoxide can be inserted into the rate law in order to predict the rate of the reaction.

The Iodination of Acetone

In today's experiment, you will be studying the kinetics of the reaction between acetone and iodine to form iodoacetone and iodide (Equation 9).

$$\underset{}{CH_3COCH_3\ (aq)} + \underset{\text{yellow}}{I_2\ (aq)} \rightarrow CH_3COCH_2I\ (aq) + H^+\ (aq) + \underset{\text{colorless}}{I^-\ (aq)} \qquad \text{Equation 9}$$

The rate law will be determined by varying the concentration of acetone and iodine. To study reaction rates it is necessary to measure the concentration of reactants as a function of time at the start of the reaction (the "initial rate" in Table 1), making kinetic studies typically difficult. In this particular experiment, however, the amount of acetone will be kept in vast excess with respect to the amount of iodine so that the concentration of acetone does not change appreciably during the course of the reaction. As a result, the rate of the reaction remains relatively constant throughout the course of the reaction. In other words, the "initial rate" that we need in order to determine the rate law will be equated with the average rate of the reaction. The equation used to find the average rate of reaction (Equation 10) is found by measuring the change in iodine concentration divided by the time needed to react.

$$\text{Rate} = \frac{-\Delta[I_2]}{\Delta t} = \frac{-([I_2]_{final} - [I_2]_{initial})}{\Delta t} \qquad \text{Equation 10}$$

The study of the reaction for the iodination of acetone is also made easy due to the color changes of the solution. Iodine (I_2) is a pale yellow whereas the iodide ion (I^-) is colorless. Hydrochloric acid, which is introduced to the reactant solution as a catalyst, is also colorless. Therefore, changes in iodine concentration can easily be visualized. The time at which the pale yellow color of the initial solution turns clear indicates that the reaction is complete and that $[I_2]_{final}$ = 0 M.

OVERVIEW

After following a procedure for the preparation of Solution 1, you will devise your own experimental protocol for creating Solutions 2,3, and 4 in order to determine the order with respect to acetone and iodine, the value of the rate constant, and predictions concerning reaction rate.

PROCEDURE: Students work in pairs

Chemicals Used	Materials Used
4 M Acetone (100 mL) 1 M HCl (100 mL) 0.00118 M Iodine (100 mL) De-ionized water (DI H_2O)	50-mL Graduated cylinders (4) 100-mL Beakers (4) 125-mL Erlenmeyer flasks (5) Plastic pipet (4) Stop-watch Stir-plate and stir bar (optional)

1. Thoroughly clean four 100-mL beakers, four 50-mL graduated cylinders and five 125-mL Erlenmeyer flasks with soap and water. Rinse all the glassware with de-ionized water and allow to dry. Label one of each beaker and graduated cylinder with the following: "acetone", "HCl", "I_2", "DI H_2O". Label four of the Erlenmeyer flasks with numbers 1 – 4 and label the last flask "blank". Rinse each beaker/graduated cylinder with 2-3 mLs of the solution named on the label. Pour approximately 50 mL of the appropriate solution into the labeled 100-mL beakers.

2. Prepare a blank that you will use for a color comparison. The blank should consist of 50-mL of water in a 125-mL Erlenmeyer flask.

3. With the appropriately labeled 50-mL graduated cylinders, add 10.0 mL of acetone, 10.0 mL of HCl and 20.0 mL of de-ionized water into a clean 125-mL Erlenmeyer flask (labeled Solution #1).

4. Measure 10.0 mL of 0.00118 M iodine into the clean "I_2" graduated cylinder. Place the Erlenmeyer flask (Solution 1) onto the stir-plate and drop a stir bar into the solution. Set the stir plate to a medium setting. Quickly pour all the iodine solution into the Erlenmeyer flask. Immediately begin timing the reaction as soon as all the iodine has been transferred to Solution #1. The solution will appear yellow due to the presence of iodine. The color will fade as the iodine reacts with the acetone. Record the time when the color of iodine just disappears by comparison with the blank. Record the volumes of acetone, HCl, iodine and H_2O for Solution #1.

5. Repeat steps 3 and 4. Calculate and record the percent difference between the two times. Repeat until the percent difference is less than 5%.

6. With your partner, decide how to alter the composition of Solution #1 to determine the order with respect to iodine (Solution #2) and the order with respect to acetone (Solution #3). The only requirement is that you must maintain a total volume of 50 mL and the volume of HCl must be 10 mL. After showing your proposal to your instructor, carry out the reactions for Solutions 2 and 3 exactly as described for Solution #1.

7. For your final reaction, devise Solution #4 using reactant volumes that you have not previously used. Remember, the total volume must remain 50 mL and that 10 mL of the total volume must be 1 M HCl. Record the time of the reaction.

Determining the Rate Law: A Kinetics Study of the Iodination of Acetone

Name:	Lab Instructor:
Date:	Lab Section:

PRE-LABORATORY EXERCISES

1. According to your textbook, what are the four factors that affect the rate of a chemical reaction? Which of these factors will be studied in this experiment?

2. Distinguish among the following terms: initial rate, average rate, and instantaneous rate.

   Which of these rates would you expect to have the largest value? Explain.

   Which of the rates are typically used to determine the rate law for a reaction?

   Which of these rates will we use to determine the rate law for a reaction?

3. Use Equation 9 to predict the initial rate if $[NO]_{initial}$ = 0.30 M and $[H_2]_{initial}$ = 0.15 M.

   What would happen to the initial rate of the reaction if $[NO]_{initial}$ = 0.60 M and $[H_2]_{initial}$ = 0.15 M instead. Does your result make sense in terms of the order of the reaction?

OVER →

PRE-LABORATORY EXERCISES continued...

4. Assuming that concentrations are expressed in moles per liter and time in seconds, what are the units of the rate constant, k, for an overall first order rate law? Show your work.

   What are the units of k for an overall second order rate law? Show your work.

   Using these two rate constants as examples, write a general rule to explain how the units of the rate constant depend on the overall order of the rate law.

5. Write the general form of the rate law for the reaction in Equation 9.

Determining the Rate Law: A Kinetics Study of the Iodination of Acetone

Name:	Lab Instructor:
Date:	Lab Section:

RESULTS and POST-LABORATORY QUESTIONS

	Solution #1	Solution #2	Solution #3	Solution #4
Volume, 4.0 M acetone				
Volume, 1 M HCl				
Volume, H_2O				
Volume, 0.00118 M iodine				
Reaction time, Trial 1 *				
Reaction time, Trial 2 *				
Average reaction time				

* For Solution #1, only record the times for the two trials within 5%.

Summarize your results as in Table 1 in the BACKGROUND:

Solution #	$[\text{acetone}]_{\text{initial}}$, M ##	$[\text{iodine}]_{\text{initial}}$, M ##	Initial rate, M/s **
1			
2			
3			
4			

Be sure to account for your dilution to 50 mLs.

** Use Equation 10 in BACKGROUND.

1. Determine the rate law (including the values for the orders of the reaction and the value for the rate constant with units) for the reaction studied in this experiment. Show all your work.

2. Use the rate law to make a prediction for the theoretical initial rate of the reaction for Solution #4. How does it compare to the experimental initial rate for Solution #4?

OVER →

RESULTS and POST-LABORATORY QUESTIONS continued...

3. Why is it important to keep the total volume of Solutions #1-4 at 50 mLs? If more water had been introduced to one of the solutions (giving a total volume of 60 mLs), would you expect the reaction rate to increase or decrease? Explain.

4. The following reaction occurs without a change in the color. $2A\ (g) + B_2\ (g) \rightarrow 2AB\ (g)$
   a) How could you monitor the concentration of the reactants and products?

   b) How would you determine the reaction orders?

   c) How would you find the rate constant and the units for the rate constant?

Determining the Equilibrium Constant of a Complex

Carrie Lopez-Couto

OBJECTIVES

- Develop a conceptual understanding of the equilibrium constant, K_c.
- Practice diluting solutions.
- Use a spectrometer to determine the value of the equilibrium constant for the formation of $FeSCN^{2+}$ (aq).

INTRODUCTION

Imagine three young students in a classroom. In this classroom there is a large chest with many toys (Figure 1). The children are constantly taking out toys without putting any back. Soon their teacher, Ms. Lopez-Couto tells the children that it is time to pick up the toys. The two girls listen to their teacher and begin to pick up the toys. The only problem is that every time the two girls each pick up a toy and returns it to the toy chest, the boy has taken two out. Ten minutes later when the teacher returns to make sure the job has been done she is shocked; there are just as many toys on the floor now as there were before. "This room is a mess! What have you been doing for the ten minutes?!" scolds the teacher. To which the girls reply, "But Ms. Lopez-Couto, haven't you ever heard of dynamic equilibrium?"

Figure 1. A toy chest provides an opportunity to study the equilibrium process.

BACKGROUND

Dynamic Chemical Equilibrium

After a given amount of time all chemical reactions appear to stop. In most cases this occurs before the reaction is "complete." The result is a mixture of products and reactants. For example, consider the system in Equation 1.

$$N_2O_4\ (g) \leftrightarrow 2NO_2\ (g) \qquad \text{Equation 1}$$

colorless brown

The progress of this reaction can be identified by noting the color change. As time passes, the gas in the closed tube turns incrementally darker as colorless gaseous N_2O_4 dissociates into brown gaseous NO_2 (Figure 2).

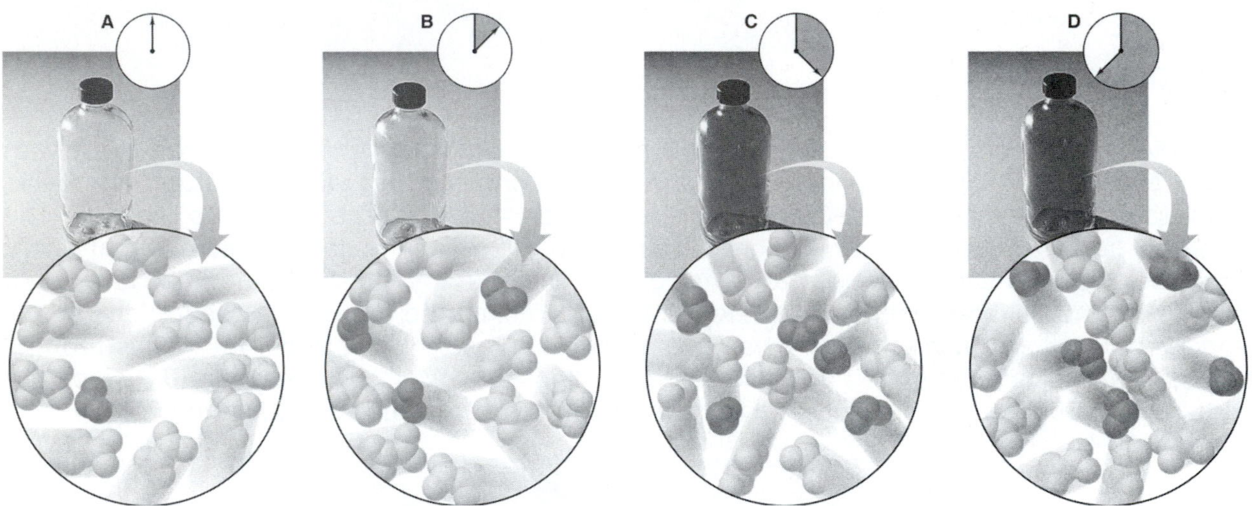

Figure 2. The generation of NO_2 is seen in the increasing brown color of the sample. Once equilibrium is reached (Time C), there is no further color change for the sample

Eventually the color change ends even though some N_2O_4 (g) remains. For instance, if you were to return an hour later no further color change would have occurred. This will most likely make you think that no reaction is taking place when, in fact, a lot of activity is occurring. The reaction is running at the same rate in opposite directions which causes the reaction to appear dormant. This condition is called a dynamic chemical equilibrium. A dynamic chemical equilibrium exists when the forward and reverse reactions are occurring at equal rates. The result is that the rate at which the products are produced from reactants is equal to the rate at which reactants are produced from products. For example, given the reaction in Equation 1, the rate laws can be written where k_f and k_r are the rate constants for the forward (Equation 2) and reverse (Equation 3) reactions respectively.

Forward reaction: $N_2O_4\ (g) \rightarrow 2NO_2\ (g)$

$$\text{rate (f)} = k_f\,[N_2O_4] \qquad \text{Equation 2}$$

Reverse reaction: $2NO_2\ (g) \rightarrow N_2O_4\ (g)$

$$\text{rate (r)} = k_r\,[NO_2]^2 \qquad \text{Equation 3}$$

Imagine we start with pure N_2O_4 (g) in a sealed container. As the N_2O_4 decomposes to form NO_2, the concentration of N_2O_4 decreases (and so does the rate in the forward direction) while the concentration of NO_2 increases (and so does the rate in the reverse direction).

Eventually the system reaches chemical equilibrium when the rate in the forward direction equals the rate in the reverse direction. At this point, Equation 2 and Equation 3 can be set equal to each other (Equation 4).

$$k_f\,[N_2O_4] = k_r\,[NO_2]^2 \qquad \text{Equation 4}$$

Equation 4 can be rearranged into a very useful form (Equation 5) that is set to equal a new constant, the equilibrium constant, K_c.

$$\frac{k_f}{k_r} = \frac{[NO_2]^2}{[N_2O_4]} = K_c \qquad \text{Equation 5}$$

Looking at Equation 5 we see that the equilibrium constant, K_c, is a ratio of the forward and reverse rate constants. The value of K_c is a constant at a given temperature because regardless of how much you start with, either product or reactant, the system eventually reaches equilibrium concentrations that when plugged into the equilibrium constant expression will give you the same constant value.

Determining the Equilibrium Constant

As an example of how to determine the value of an equilibrium constant, imagine you are a scientist and you need to determine K_c for the production of ammonia gas from gaseous hydrogen and nitrogen at a temperature of 1000 K (Equation 6). When the K_c expression is written, the coefficients in the balanced reaction become subscripts and pure solids or liquids are not included. The equilibrium constant expression for this reaction is shown in Equation 7.

$$N_2\,(g) + 3H_2\,(g) \leftrightarrow 2NH_3\,(g) \qquad \text{Equation 6}$$

$$K_c = \frac{[NH_3]^2}{[N_2][H_2]^3} \qquad \text{Equation 7}$$

Finding K_c is as simple as plugging in the equilibrium values of each species. Unfortunately, finding these equilibrium concentrations can take some work. So, you set up a reaction vessel with the following initial concentrations: $[N_2]$ = 0.500 M, $[H_2]$ = 0.500 M, and $[NH_3]$ = 0.000 M. After the system has reached equilibrium at the desired temperature, you use an NH_3 sensor and find that the concentration of NH_3 is 0.012 M. The other values are calculated by using what is called an **I**nitial/**C**hange/**E**quilibrium table or ICE table (Table 1).

Begin by setting up the table and filling in any known values. In this case, we know all of the "**I**nitial" concentrations. We don't know what "**C**hange" the system undergoes, but we do know that for every N_2 molecule that is used up, three H_2 molecules are also used up and two NH_3 molecules are made. Based on stoichiometry, we can fill in the variables (x's) for the "**C**hange" in Table 1. The "**E**quilibrium" values are then taken as the sum of the "**I**nitial" values and the "**C**hange".

Table 1. An ICE table for determining K_c of Equation 6

	$[N_2]$ (M)	$[H_2]$ (M)	$[NH_3]$ (M)
Initial	0.500	0.500	0.000
Change	-x	-3x	+2x
Equilibrium	0.500 – x	0.500 – 3x	2x

Since we know that $[NH_3]_{equilibrium}$ = 0.012 M = 2x, we know that x = 0.0060 and we can solve for the other "**E**quilibrium" values in Table 1. We get $[N_2]_{equilibrium}$ = 0.494 M and $[H_2]_{equilibrium}$ = 0.482 M. We now plug the equilibrium concentrations into Equation 7, giving an equilibrium constant for the production of ammonia at this temperature of 2.6×10^{-3}. The magnitude of the equilibrium constant indicates the equilibrium position for the chemical system. The larger the value of the equilibrium constant, the farther the equilibrium lies to the right. The small value of K_c for our reaction indicates that at equilibrium the reactants are the predominate species.

Today's Laboratory Experiment

In this experiment you'll determine the equilibrium constant for the reaction in Equation 8.

$$Fe^{3+}(aq) + SCN^{-}(aq) \leftrightarrow \underset{\text{red}}{FeSCN^{2+}(aq)} \qquad \text{Equation 8}$$

Since the product, $FeSCN^{2+}$ is a reddish complex ion with an absorption maximum (λ_{max}) at 447 nm, its concentration can be determined spectrophotometrically after generating a calibration curve for absorbance versus $[FeSCN^{2+}]$. Your calibration curve for this experiment will consist of only one data point corresponding to the absorbance of Solution A. Solution A is the only solution you will make where you know the value of $[FeSCN^{2+}]$ because in solution A, we have Fe^{3+} in enough excess to ensure that all of the SCN^- has been converted to $FeSCN^{2+}$. Your calibration curve will therefore consist of a line connecting the origin with point A.

OVERVIEW

In this experiment you will analyze several different solutions containing the $FeSCN^{2+}$ complex. Shown below are some steps that will help you to calculate the equilibrium constant.

- Carefully prepare Solutions A, B and C according to directions.
- Measure % transmittance of Solutions A, B and C using a spectrophotometer.
- Convert % transmittance values to absorbance, where $A = -\log\left(\frac{\%T}{100}\right)$
- Plot a calibration curve using points (0,0) and the absorbance values for solution A.
- Find the slope of this line.
- Using the slope of the line, calculate the equilibrium concentration of $FeSCN^{2+}$ in Solutions B and C.
- Use an ICE table to determine the equilibrium concentration of Fe^{3+} and SCN^- in solutions B and C.
- Calculate the experimental value of equilibrium constant for Solutions B and C.

PROCEDURE (Students work in pairs, Class data will be analyzed)

Chemicals Used	Equipment Used
2.0×10^{-4} M stock KSCN	10-mL pipet
5.0×10^{-2} M stock $Fe(NO_3)_3$ solution (#1)	50-mL Erlenmeyer flask (3)
8.0×10^{-3} M stock $Fe(NO_3)_3$ solution (#2)	100-mL beaker
3.2×10^{-3} M stock $Fe(NO_3)_3$ solution (#3)	

1. Turn on the spectrometer (Figure 3) by rotating the power control clockwise. Allow the spectrometer to warm-up for five minutes before using.

2. Clean a 10-mL pipet with very dilute soap and water. Rinse the pipet first with deionized water and then several times with a few milliliters of KSCN solution. Dispose of the waste KSCN solution and all subsequent rinsing solutions in a 100-mL waste beaker.

3. Clean and dry three 50-mL Erlenmeyer flasks. Label them A, B and C. Pipet 10.00-mL portions of KSCN stock solution into each of the three labeled 50-mL Erlynmeyer flasks.

4. Rinse the pipet with deionized water and then with with several milliliters of the 0.050 M stock $Fe(NO_3)_3$ solution (#1). Add 10.00 mLs of 5.0×10^{-2} M stock $Fe(NO_3)_3$ solution (#1) to flask A.

5. Repeat step 4, by rinsing the pipet and adding 10.00 mLs of 8.0×10^{-3} M stock $Fe(NO_3)_3$ solution (#2) to flask B and 3.2×10^{-3} M stock $Fe(NO_3)_3$ solution (#3) to flask C.

6. Adjust the wavelength on the spectrometer to 450 nm. With no sample in the spectrometer, turn the zero adjust so the meter reads 0% T. Three-quarters fill a clean cuvette with de-ionized water (blank solution). Insert the cuvette into the sample holder of the spectrometer and adjust the light-control knob so 100% transmittance is read. Your instrument is now zeroed. Measure the % transmittance of solutions A, B and C at 450 nm. Remember, whenever you are using a new sample, to rinse the cuvette with a few mLs of the solution.

7. On the board with the class data, write your calculated values for the equilibrium constant, K_c, for solutions B and C. Record all of the class data so you can determine the class average. Dispose of the waste beaker contents according to instructor's directions.

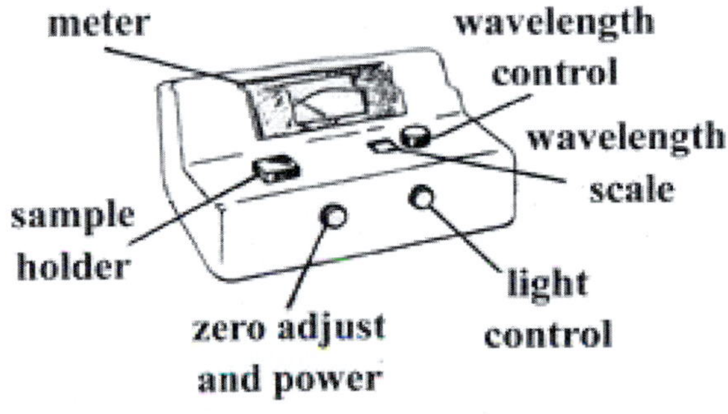

Figure 3. A typical spectrometer

Determining the Equilibrium Constant of a Complex

Name:	Lab Instructor:
Date:	Lab Partner:

PRE-LABORATORY EXCERCISES

1. You dilute 0.250 L of a 0.80 M NaOH stock solution to 1.20 L. What is the new concentration of the NaOH?

2. Given a reaction with a measured equilibrium constant, K_c, of 1 x 10^{-30}, what can be said about the equilibrium position?

3. Write the equilibrium constant expressions for each of the following reactions:
   a) $2NOBr\ (g) \leftrightarrow 2NO\ (g) + Br_2\ (g)$

   b) $Ag^+\ (aq) + Fe^{2+}\ (aq) \leftrightarrow Ag\ (s) + Fe^{3+}\ (aq)$

   c) The reaction in Equation 8.

4. An empty vessel is charged with 1.5 atm of SO_3 gas. The system reaches equilibrium according to the following reaction: $2SO_3\ (g) \leftrightarrow 2SO_2\ (g) + O_2\ (g)$. After equilibrium was reached, the partial pressure of SO_3 was found to be 1.1 atm. Using an ICE table, calculate the value of K_p for this system at this particular temperature.

Determining the Equilibrium Constant of a Complex

Name:	Lab Instructor:
Date:	Lab Partner:

RESULTS and POST-LABORATORY QUESTIONS

Stock solution concentrations

Concentration of SCN^- in stock solution ____________

Concentration of Fe^{3+} in stock solution #1 ____________

Concentration of Fe^{3+} in stock solution #2 ____________

Concentration of Fe^{3+} in stock solution #3 ____________

Create a calibration curve

Solution	% T at 450 nm	A at 450 nm	$[FeSCN^{2+}]_{equil}$ **
A			

**assume all SCN^- in sample A is converted to $FeSCN^{2+}$

Attach a copy of your calibration curve using solution A. (absorbance vs. $[FeSCN^-]$)

What is the slope of the line you obtained from the calibration curve?

From your calibration curve (use the slope of the line) and the absorbance values, calculate the equilibrium concentration of $FeSCN^{2+}$ for solutions B and C. Record your data below.

Solution	% T at 450 nm	A at 450 nm	$[FeSCN^{2+}]_{equil}$
B			
C			

Using the initial concentrations of Fe^{3+} and SCN^- and the amount that was lost while making the equilibrium amount of $FeSCN^{2+}$, calculate the equilibrium concentrations of Fe^{2+} and SCN^- in solutions B and C (this should be done with an ICE table). Show all of your work for these calculations and then fill these values in the table below.

Solution	$[Fe^{3+}]_{initial}$	$[Fe^{3+}]_{equil}$	$[SCN^-]_{initial}$	$[SCN^-]_{equil}$
B				
C				

OVER →

RESULTS and POST-LABORATORY QUESTIONS continued...

Show all your work for the calculation of K_c for solutions B and C.

What are the average class values for K_c for the two solutions? Are the relative values for the two solutions what you expected? Explain.

Le Chatelier's Principle: "Stress Management"

Carrie Lopez-Couto

OBJECTIVES

- Develop a conceptual understanding of Le Chatelier's Principle.
- Determine the effect of concentration and temperature changes on a chemical system at equilibrium.

INTRODUCTION

Le Chatelier's Principle states that if an external stress is applied to a chemical system at equilibrium, the equilibrium will shift in the direction that minimizes the effect of the stress. In fact, Le Chatelier's Principle can be applied to any system at equilibrium. For example, Le Chatelier's Principle describes the delicate equilibrium that is maintained between the population of herbivores (antelope, wildebeest, zebra) and carnivores (lion, cheetah) on the African Savannah (Figure 1). When there is a stress or disturbance on the equilibrium, such as a disease or a drought, a decrease in the number of herbivores will cause a shift in all the populations until a new equilibrium is attained.

Figure 1. Carnivores and herbivores on the African Savannah exist in a carefully balanced state of equilibrium

BACKGROUND

Le Chatelier's Principle

As stated in the introduction, when an external stress is applied to a system at equilibrium, the system responds by shifting its equilibrium to accommodate the stress. In terms of chemical systems there are three primary stresses that you should be familiar with: changes in concentration, changes in temperature and changes in pressure. This laboratory experiment will focus on the first two.

The Effect of Concentration

To explore the effect of concentration on a system at equilibrium, we will study the chromate (CrO_4^{2-})/dichromate ($Cr_2O_7^{2-}$) equilibrium described in Equation 1.

$$\underset{\text{yellow}}{2\ CrO_4^{2-}}\ (aq) + 2\ H^+\ (aq) \leftrightarrow \underset{\text{orange}}{Cr_2O_7^{2-}}\ (aq) + H_2O\ (l) \qquad \text{Equation 1}$$

The key to distinguishing between chromate and dichromate is their color. The chromate is a bright yellow and the dichromate is orange. The color of the solution will help you to determine the dominant species present in the solution. An orange solution, for example, indicates that the solution contains predominately the products indicated in Equation 1.

Imagine you start with a test-tube of water and add a few grams of CrO_4^{2-} (maybe in the form of Na_2CrO_4) and a few drops of acid. The system will eventually reach equilibrium and the color will stabilize. Depending on the amount of acid added and the temperature, the position of the equilibrium will vary and the solution may be yellow, orange or somewhere in between. Le Chatelier's Principle predicts that if we then increase the H^+ concentration (by adding more acid) the system will shift in the direction that will decrease, or counter balance, the change made (the common ion effect). In other words, we predict the reaction will use up some of the added H^+ by shifting toward the products. Whether or not this prediction holds true can be determined by looking for the color change. The chemical basis for this prediction relates to the kinetics of the reaction. Increasing the concentration of a substance causes a temporary increase in the reaction rate on that side of the equilibrium. Upon addition of acid, the hydrogen ion concentration increases, and as a result, the chromate and hydrogen ions are converted to products faster than the reverse process. This shift in the equilibrium continues until the system gradually reaches new equilibrium concentrations.

This is confirmed by referring to the equilibrium constant, K_c, for our reaction (Equation 2).

$$K_c = \frac{[Cr_2O_7^{2-}]}{[H^+]^2[CrO_4^{2-}]^2} \qquad \text{Equation 2}$$

By increasing the concentration of H^+, the denominator in Equation 2 increases. As a result the value of the reaction quotient (Q) is less than K_c and the reaction shifts to the right in order to reestablish the value of K_c. Eventually the concentration of H^+ and CrO_4^{2-} will decrease enough and the concentration of $Cr_2O_7^{2-}$ will increase enough that the value of K_c is restored (even though the concentrations have all changed).

The Effect of Temperature

Temperature, like concentration, can also influence a system at equilibrium. When determining the change in an equilibrium position due to changes in temperature, it is necessary to determine whether the reaction is endothermic or exothermic. Consider the reaction between gaseous phosphorus trichloride and gaseous chlorine in Equation 3.

$PCl_3\ (g) + Cl_2\ (g) \leftrightarrow PCl_5\ (g)$ $\Delta H^\circ_{rxn} = -111$ kJ Equation 3

Based on the sign of ΔH°_{rxn} we know that the forward reaction is exothermic and the reverse reaction is endothermic. We can therefore rewrite Equation 3 as shown in Equation 4:

$PCl_3\ (g) + Cl_2\ (g) \leftrightarrow PCl_5\ (g) + \text{heat}$ Equation 4

If the temperature of the system is then increased, the system responds by absorbing the heat added. This is achieved by favoring the endothermic direction (in this case, the reverse reaction) causing some PCl_5 to decompose into PCl_3 and Cl_2. These new equilibrium concentrations result in a new value for the equilibrium constant, K_C. Again, this can be understood by examining the equilibrium constant for the reaction (Equation 5).

$$K_c = \frac{[PCl_5]}{[PCl_3][Cl_2]}$$ Equation 5

If the concentrations of PCl_3 and Cl_2 increase and the concentration of PCl_5 decreases, the new value of the equilibrium constant, K_C', will be smaller than the value before the system was heated. A change in temperature is the only disturbance that will result in a different K_C value.

The system used to test the effect of temperature in this laboratory experiment is shown in Equation 6.

$$\underset{\text{pink}}{Co(H_2O)_6^{2+}}\ (aq) + 4\ Cl^-\ (aq) \leftrightarrow \underset{\text{violet}}{CoCl_4^{2-}}\ (aq) + 6\ H_2O\ (l)$$ Equation 6

Again, the color will be used to identify the predominant species present. The hexaquocobaltate (II) complex is pink and the tetrachlorocobaltate (II) complex is violet.

OVERVIEW

In this experiment you will study the effect of concentration on the chromate-dichromate equilibrium and the effect of temperature on the cobalt chloride hexahydrate equilibrium.

PROCEDURE

Part A. Effect of Concentration (Equation 1 in BACKGROUND)

Chemicals Used	Materials Used
1.0 M HCl 1.0 M NaOH 0.10 M CrO_4^{2-} 0.10 M $Cr_2O_7^{2-}$	24-well plate Plastic disposable pipets

Caution: Solutions of HCl and NaOH are caustic and should be handled with care. If you spill any chemicals on your skin, wash with running water and inform your instructor.

1. Label four wells on a 24-well plate. Add 10 drops of CrO_4^{2-} (aq) to wells #1 and #3 and 10 drops of $Cr_2O_7^{2-}$ (aq) to wells #2 and #4. Record the color of each solution.

2. Add 5 drops of 1.0 M HCl to wells #1 and #2 and 5 drops of 1.0 M NaOH to wells #3 and #4. Record the color of each solution.

3. Add 10 drops of 1.0 M NaOH to well #1 and 10 drops of 1.0 M HCl to well #3. Record the color of each solution.

Part B. Effect of Concentration and Temperature (Equation 6 in BACKGROUND)

Chemicals Used	Materials Used
0.1 M $CoCl_2$ 6 M HCl $CaCl_2$ (s) 0.1 M $AgNO_3$ Ice	24-well plate Disposable pipets Large test tube Test tube holder Hot plate 250-mL Beakers for hot/cold water baths (2)

Caution: Solutions of HCl are caustic and should be handled with care. If you spill any chemicals on your skin, wash with running water and inform your instructor.

Using a procedure similar to the one in Part A, devise a protocol for exploring the effects of adding 6 M HCl, solid $CaCl_2$ and 0.1 M $AgNO_3$ to 0.1 M $CoCl_2$. After showing your instructor, carry out your plans. Record all observations.

Effect of temperature: Using a pipet, half-fill a clean test tube with 2.0 mL of 0.1 M $CoCl_2$. Record your observations after performing each of the following steps: Place the test tube into a boiling water bath for 5 minutes. Place the test tube into an ice-bath for 5 minutes.

Le Chatelier's Principle: " Stress Management"

Name:	Lab Instructor:
Date:	Lab Section:

PRE-LABORATORY EXERCISES

1. Define the underlined terms in the BACKGROUND section.

2. Describe a non-chemical application of Le Chatelier's principle. What is the system in equilibrium? What is the stress? How does the system respond?

3. Consider the equilibrium discussed in Equation 3 in the BACKGROUND. As head chemist at a chemical plant, you are in charge of the amount of PCl_5 that is produced daily. List two ways to produce more PCl_5 (one involving changing temperature and one involving changing the concentration).

Le Chatelier's Principle: " Stress Management"

Name:	Lab Instructor:
Date:	Lab Section:

RESULTS and POST-LABORATORY QUESTIONS

Part A. Effect of Concentration (Equation 1 in BACKGROUND)

In the table below, fill in the color and the predominate species present in each well after each of the steps in the procedure.

	Well #1	Well #2	Well #3	Well #4
Step 1: add solutions				
Step 2: add HCl				
Step 3: add NaOH				

Do the color changes indicate that a physical or a chemical change has occurred? Explain.

How does Le Chatelier's principle explain the results you obtained after the addition of HCl to well #1 (Step 2)?

How does Le Chatelier's principle explain the results you obtained after the addition of NaOH to well #1 (Step 3)?

Concentrated H_2SO_4 is a dehydrating agent. If after Step 3, you continued by adding H_2SO_4 to wells #1-4 what would you expect to happen to each of the colors? Explain.

OVER →

RESULTS and POST-LABORATORY QUESTIONS continued...

Part B. Effect of Concentration and Temperature (Equation 6 in BACKGROUND)

Make and attach a table similar to the one provided for you in Part A. In addition to listing the color and predominate species present in each well, be sure to list what you did in each step.

What ions are present in a solution of HCl? Which of these ions will effect the equilibrium of $CoCl_2$? Does this agree with your observations? Support your answer using references to specific wells and steps from your table.

How does Le Chatelier's principle explain the results you obtained after the addition of solid $CaCl_2$ to $CoCl_2$.

What ions are present in a solution of $AgNO_3$? Which of these ions will effect the equilibrium in Equation 6? Explain. Does this agree with your observations? Support your answer using references to specific wells and steps from your table.

Is the reaction in Equation 6 exothermic or endothermic? Explain using your experimental data from Part B.

Chemistry of the Kitchen: Acids and Bases

D. Van Dinh

OBJECTIVES

- Classify common household chemicals as either acids or bases using a homemade indicator solution.
- Develop an understanding of the pH scale.
- Examine the differences between strong and weak acids.
- Titrate a sample of vinegar to determine the concentration of acid.

INTRODUCTION

Take a moment and look around your house. Many everyday chemicals used in your home are acids or bases. Some of the most common acids and bases you may recognize and use include vinegar (acetic acid), lemon juice (citric acid), and ammonia. Acids and bases are essential substances in home, industry, and the environment. For example, the vast quantity of sulfuric acid manufactured in the United States each year is needed to produce fertilizers, polymers, steel, and many other materials. The influence of acids on the environment can be seen through acid rain, which has caused numerous historic buildings and monuments to erode. If you have ever had a goldfish, you know how important it is to monitor and control the acidity of the water in the aquarium (Figure 1). A characteristic that acids and bases share is their ability to turn certain organic compounds, such as vegetable materials, distinctive colors. Utilizing this knowledge, what are some methods in which you could test the acidity of the water in your goldfish aquarium?

Figure 1. A healthy fish tank requires careful control of acidity levels.

BACKGROUND

Properties of Acids and Bases

Since the 17th century, acids and bases have been characterized by their sour and bitter tastes, respectively. In modern chemistry, these concepts have taken on considerably more precise meaning. In fact, there are three definitions of acids and bases, the classical (Arrhenius), the Bronsted-Lowry, and the Lewis, which greatly expand our knowledge of these chemicals. Acids and bases differ greatly in their strength in water, that is, in the amount of H_3O^+ or OH^- produced per mole of substance dissolved. They are generally classified as either strong or weak, depending on the extent of their dissociation into ions in water. Acids and bases are electrolytes in water, so the classification of acid and base strength correlates with the classification of electrolyte strength: strong electrolytes such as the strong acids dissociate completely (Figure 2a), and weak electrolytes, such as the weak acids undergo partial dissociation (Figure 2b).

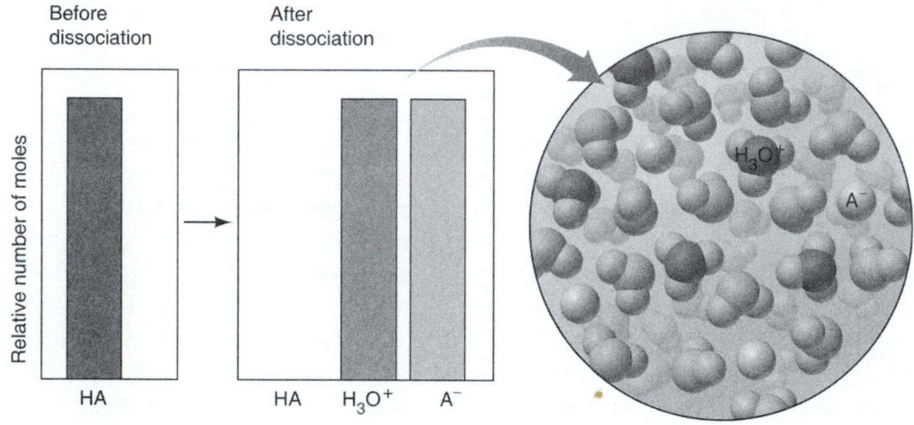

Figure 2a. The behavior of a strong acid in water.

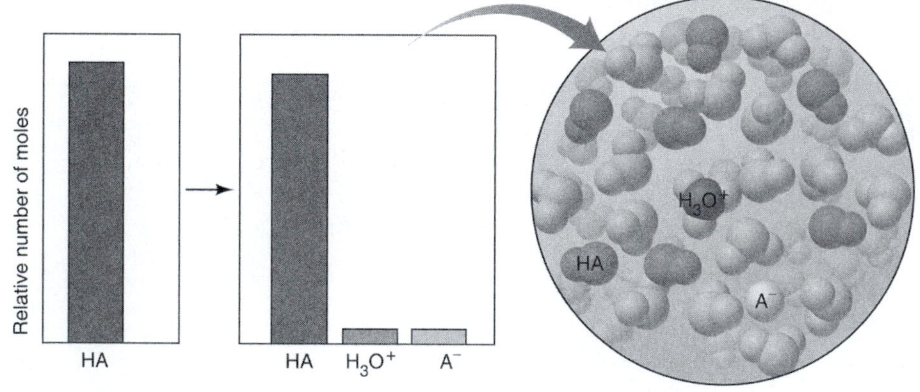

Figure 2b. The behavior of a weak acid in water.

In a solution of a strong acid, virtually no HA molecules are present (Equation 1).

$$HA\ (aq) + H_2O\ (l) \rightarrow H_3O^+\ (aq) + A^-\ (aq) \qquad \text{Equation 1}$$

The $[H_3O^+] = [A^-] \approx [HA]_{initial}$ and the $[HA]_{equilibrium} \approx 0$. Since this process essentially goes to completion, a single arrow is used ($\rightarrow$) in the case of strong acids.

The situation is different in the case of a weak acid. From Figure 2b, the majority of HA molecules in a solution of a weak acid are undissociated. Thus, $[H_3O^+] \ll [HA]_{initial}$ and $[HA]_{equilibrium} \approx [HA]_{initial}$. Notice in Equation 2 that the reaction is expressed as an equilibrium ($\leftrightarrow$) indicating that the reaction does not necessarily produce 100% products. In fact, in the case of most weak acids, less than 5% of the original HA molecules will actually dissociate to produce H_3O^+ *(aq)* and A^- *(aq)* ions.

$$HA\ (aq) + H_2O\ (l) \leftrightarrow H_3O^+\ (aq) + A^-\ (aq) \qquad \text{Equation 2}$$

The percentage of acid that is actually dissociated can be quantified in terms of the acid dissociation constant, K_a, whose expression for the dissociation of a general weak acid, HA, in water is shown in Equation 3. Notice $[H_2O]$ does not show up in the form of K_a.

$$K_a = \frac{[H_3O^+][A^-]}{[HA]} \qquad \text{Equation 3}$$

Like any equilibrium constant, the magnitude of K_a tells how far to the right the reaction has proceeded when equilibrium is reached. Thus, the stronger the acid, the higher the $[H_3O^+]$ at equilibrium, the larger the K_a:

Stronger acid $\Rightarrow$ higher % HA dissociation $\Rightarrow$ higher $[H_3O^+]$ $\Rightarrow$ larger K_a

Likewise, the weaker the acid, the smaller the K_a:

Weaker acid $\Rightarrow$ lower % HA dissociated $\Rightarrow$ lower $[H_3O^+]$ $\Rightarrow$ Smaller K_a

The pH Scale

In aqueous solutions, $[H_3O^+]$ can vary over an enormous range: from about 10 M to 10^{-15} M. To handle numbers with negative exponents more conveniently in calculations, we convert them to positive numbers using a numerical system called a p-scale, the negative of the common (base-10) logarithm of the number. Applying this numerical system to $[H_3O^+]$ gives pH, the negative logarithm of $[H_3O^+]$ or $[H^+]$ (Equation 4).

$$pH = -\log [H_3O^+] \qquad \text{Equation 4}$$

For example, a solution with a $[H_3O^+]$ of 5.4×10^{-4} M has a pH of 3.27 (Equation 4a).

$$pH = -\log [H_3O^+] = -\log (5.4 \times 10^{-4}) = 3.27 \qquad \text{Equation 4a}$$

Likewise, the pH of a 1.0×10^{-12} M H_3O^+ solution is 12.00 on the pH scale. Note that *the higher the pH, the lower the $[H_3O^+]$*. Therefore, an acidic solution has a lower pH (higher $[H_3O^+]$) than a basic solution. At 25°C the $[H_3O^+]$ in pure water is 1.0×10^{-7} M, so the pH of pure water at 25°C is 7.00 and aqueous solutions typically fall within a range of 0 to 14. This information is summarized in Table 1.

Type of solution	pH
basic	> 7.00
neutral	= 7.00
acidic	< 7.00

Table 1

In the laboratory, pH values are usually obtained with an acid-base indicator or with a more precise instrument called a pH meter.

Calculating the pH

It is possible using what we have learned so far to calculate the theoretical pH of any aqueous acidic solution without using a pH meter. In the case of a strong acid, our job is simple since we can assume that the concentration of $[H_3O^+]$ is the same as the starting concentration of the strong acid and we simply use the calculation from Equation 4. Weak acids, are a different story. However, given the K_a and $[HA]_{initial}$ of the weak acid, we can still calculate the pH. As an example, we will calculate the pH of a 0.10 M solution of propanoic acid (C_2H_5COOH, which we simplify as HPr). A good first step in doing these problems is to look up the value of the K_a. In the case of propanoic acid, $K_a = 1.3 \times 10^{-5}$.

In order to determine the $[H_3O^+]$ of this solution, we must next write the balanced equation and the K_a expression, as in Equations 2 (Equation 5) and 3 (Equation 5a), respectively.

$$HPr\ (aq) + H_2O\ (l) \leftrightarrow H_3O^+\ (aq) + Pr^-\ (aq) \qquad \text{Equation 5}$$

$$K_a = 1.3 \times 10^{-5} = \frac{[H_3O^+][Pr^-]}{[HPr]} \qquad \text{Equation 5a}$$

From Equation 5, we see for each mole of HPr that dissociates, one mole of H_3O^+ and Pr^- each are produced. Knowing this and given $[HPr]_{initial}$ (0.10 M) we can set up a reaction table, commonly called an **I.C.E.** table (Table 2).

	HPr *(aq)*	+ H_2O *(l)*	↔	H_3O^+ *(aq)*	+ Pr^- *(aq)*
Initial	0.10 M	-		0	0
Change	- x	-		+ x	+ x
Equilibrium	0.10 M - x	-		x	x

Table 2

In the "**I**nitial" row we include the starting concentration of the weak acid. Note that the concentration of H_3O^+ and the conjugate base, Pr^- are both initially equal to zero. Because we don't yet know how much of the HPr dissociates, we express the changes in terms of a variable, x. We know that however much HPr is lost (-x) in the "**C**hange" row, we must gain the same amount (+x) of H_3O^+ and Pr^-. The "**E**quilibrium" row is the sum of the "**I**nitial" and "**C**hange" rows.

We are now ready to substitute information from our "**E**quilibrium" row into the K_a expression and solve for x (Equation 5b).

$$K_a = \frac{[H_3O^+][Pr^-]}{[HPr]} = 1.3 \times 10^{-5} = \frac{(x)\,(x)}{(0.10 - x)} \qquad \text{Equation 5b}$$

Since K_a is small for HPr, it dissociates very little so the value of x is negligibly small compared to 0.10 M; therefore, we avoid solving the quadratic equation by approximating 0.10 M – x ≈ 0.10 M (Equation 5c).

$$1.3 \times 10^{-5} \approx \frac{(x)\,(x)}{0.10} \qquad \text{Equation 5c}$$

$$x \approx \sqrt{(0.10)\,(1.3 \times 10^{-5})} = 1.1 \times 10^{-3}\ M = [H_3O^+] \qquad \text{Equation 5d}$$

Equation 6 shows that now that we have found the $[H_3O^+]$ in our solution of 0.10 M propanoic acid, we can calculate its pH using Equation 4.

$$pH = -\log [H_3O^+] = -\log (1.1 \times 10^{-3}) = 2.96 \qquad \text{Equation 6}$$

This mathematical process can also be reversed. For example, an environmental chemist can measure the experimental pH and use it to determine the concentration of acid in a sample of acid rain.

Titration

In addition to the method described above, the concentration of an acid can also be determined experimentally by an acid-base titration. In an acid-base titration, a measured volume of acidic solution of unknown concentration is placed in a flask beneath a buret containing the known (standardized) base solution (Figure 3). A few drops of acid-base indicator, usually phenolphthalein, are added to the flask containing the acidic solution. The standardized solution of base is then added slowly to the flask until the end point is reached. The end point of the titration occurs when the indicator changes color permanently due to the presence of excess OH^- ions (phenolphthalein is colorless in acid and pink in base). Knowing the stoichiometry of the acid-base reaction and the amount of base used to reach the end point, scientists can discover the unknown $[H_3O^+]$ of the acidic solution.

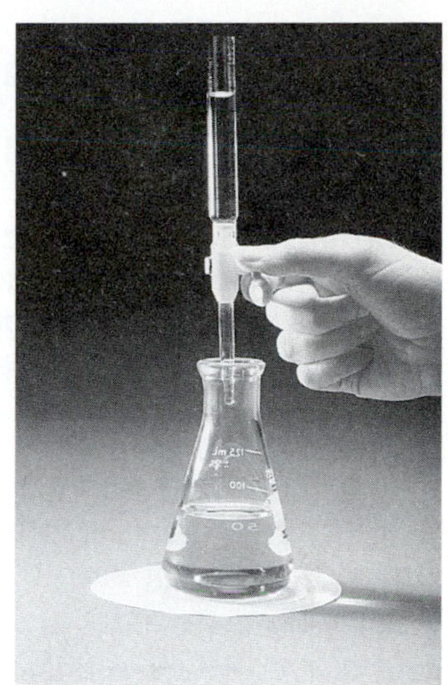

Figure 3. The experimental setup for an acid-base titration.

OVERVIEW

In this series of experiments, you will examine several of the various methods in which the pH of a solution can be measured. Knowing that acids and bases possess the ability to change the color of certain indicator solutions, you will use red cabbage extract and a universal indicator in Part A as acid-base indicators. You will develop a pH scale based on the fact that red cabbage extract changes colors when mixed with different household chemicals. In Part B, you will study the differences between strong and weak acids using a pH meter. Using this knowledge, you will then determine an unknown concentration of a weak acid. In Part C, you will calculate the concentration of acid in vinegar by performing an acid-base titration.

PROCEDURE

Part A. Relative Acidity/Basicity of Common Household Products

Chemicals Used	Materials Used
Universal indicator papers Red cabbage (or provided Red cabbage extract) Various household chemicals (ex. Ammonia, Vinegar, Shampoo (colorless), Soda (a colorless variety), Milk, Lemon juice, Liquid detergent (without dyes), Milk of magnesia, Tap H_2O, De-ionized H_2O, Bleach)	Several beakers of various sizes 10-mL Graduated cylinder Glass stirring rod Test tube w/ rubber stopper (2) Test tube rack Funnel Hot plate Knife

1. Groups of 8 – 10 students will share making a batch of cabbage juice indicator. Shred 1/4 head of cabbage and boil it in 250 mL of water for about 5 minutes. Let the mixture cool and decant off the liquid/indicator. Alternatively, your instructor may provide you with pre-made cabbage juice indicator.

2. Working alone, select one or two household products to test from the available samples. Pour about 10 mL of your selected household chemical(s) into a small beaker(s). Test the pH of your sample(s) with a strip of universal indicator paper. Let the indicator paper(s) dry and label it with the sample that you tested.

3. Using a funnel, pour your household product into a test tube. Add about 1.0 mL of red cabbage juice extract to the test tube. Stopper the tube and invert it several times. Record the color of the mixture. Repeat this procedure if you did a second household product.

4. Label your samples and bring the test tube and indicator paper to the front of the lab for observation by the class. Be sure to record results (color of indicator paper and corresponding pH from the indicator paper box as well as the color of the cabbage juice solution) for other household products.

Part B. Strong and Weak Acids

Chemicals Used	Materials Used
0.50 M HCl 0.50 M $HC_2H_3O_2$ unknown $HC_2H_3O_2$ solutions Zinc wire or shavings	pH Meter Well plate

1. At a designated area, your instructor will have labeled 100-mL beakers containing 50.0 mL of each of the chemicals listed above. Carefully following your instructor's directions, use the pH meter to record the pH of the 0.50 M HCl solution, the 0.50 M $HC_2H_3O_2$ solution and one of the unknown $HC_2H_3O_2$ solutions.

2. You instructor will have several well plates set up near the beakers with HCl and $HC_2H_3O_2$. Place a few zinc shavings or a short length of zinc wire in the bottom of two of the empty wells. Use a disposable pipet to add enough 0.5 M HCl to one of the wells to just cover the zinc. Add the same amount of 0.5 M $HC_2H_3O_2$ to the other well with the zinc. Record your observations.

Part C. Acid-Base Titration of Vinegar

Chemicals Used	Materials Used
0.20 M NaOH Vinegar Phenolphthalein	50-mL Buret and buret stand 125-mL Erlenmeyer flask (2) 100-mL Beaker (2) 5-mL Pipet w/bulb 50-mL Graduated cylinder Funnel

1. Run about 25.0 mL of de-ionized water through a 50-mL buret and then rinse the buret with about 5.0 mL of 0.20 M NaOH solution. Mount the buret in the buret stand.

2. Use a funnel to add roughly 25.0 mL of the 0.20 M NaOH solution into the buret. Drain any air bubbles from the tip of the buret into a waste beaker. Record the initial volume (± 0.02 mL).

3. Pipet 5.00 mL of the vinegar into a 125-mL Erlenmeyer flask. Add 25.0 mL of de-ionized H_2O and 2-3 drops of phenolphthalein to the flask.

4. Slowly add the 0.20 M NaOH solution drop-wise from the buret to the vinegar, swirling the flask after each addition. Continue until the endpoint is reached. Record the final volume (± 0.02 mL).

5. Repeat steps 2 - 4 for a second trial.

Chemistry of the Kitchen: Acids and Bases

Name:	Lab Instructor:
Date:	Lab Section:

PRE-LABORATORY EXERCISES

1. Define the <u>underlined</u> words in the **BACKGROUND** section.

2. Determine the concentration of H_3O^+ in a solution with a pH of 9.78.

3. Given solutions of the same concentration, which would you expect to have a lower pH, a strong or weak acid? Explain.

4. In this laboratory experiment you will be calculating the concentration of an unknown sample of acetic acid using the pH of the sample and the K_a for acetic acid. Use your textbook to find the K_a for acetic acid. Be sure to record this value in your notebook so you will have it available during the experiment.

OVER →

PRE-LABORATORY EXERCISES continued...

5. You will be performing a titration of vinegar (an aqueous solution of acetic acid, $HC_2H_3O_2$) in this laboratory experiment. To prepare you for this titration, please read the section on acid-base titrations in your textbook and then do the following:
   a) Write the balanced molecular equation and the net ionic equation for the neutralization reaction between aqueous acetic acid and aqueous sodium hydroxide.

   b) You place 10.00 mLs of a $HC_2H_3O_2$ solution of unknown concentration in a flask and add a few drops of indicator. You then titrate the acid with 0.24 M NaOH. If the initial reading on the buret was 0.19 mL and the final reading was 26.5 mLs, what is the concentration of the $HC_2H_3O_2$ solution?

   c) Some solutions (such as vinegar) are commonly reported in terms of percent by mass. Assuming the density of the acetic acid solution you found in question 6b is the same as the density of pure water at 25°C, determine the percent by mass of the vinegar in the sample.

Chemistry of the Kitchen: Acids and Bases

Name:	Lab Instructor:
Date:	Lab Section:

RESULTS and POST-LABORATORY QUESTIONS

Part A. Relative Acidity/Basicity of Common Household Products
Attach a table summarizing the colors of the universal indicator paper and the cabbage juice indicator for each of the samples that your class tested. Put a star next to the household product(s) that you personally tested. Your table should include a column that correlates the color of the universal indicator paper with the pH of the solution (this information is found on the box of the universal indicator paper).

Use the information in your table to make a new chart that would allow someone using only the cabbage juice indicator to determine the pH of a new substance.

Part B. Strong and Weak Acids
What did you observe for the reactivity of zinc metal with a weak acid and a strong acid of the same concentration? Do these observations make sense? Explain.

	Measured pH	Theoretical pH (show work below *)	% error
0.50 M HCl			
0.50 M $HC_2H_3O_2$			

*Show your work for your calculation of the theoretical pH of the 0.50 M HCl solution.

*Show your work for your calculation of the theoretical pH of the 0.50 M $HC_2H_3O_2$ solution.

OVER →

RESULTS and POST-LABORATORY QUESTIONS continued...

Part B. Strong and Weak Acids continued...

	Unknown code	Measured pH	Theoretical Concentration*
Unknown $HC_2H_3O_2$			

*Show your work for the calculation of the concentration of your $HC_2H_3O_2$ unknown based on the pH of the sample.

Part C. Acid-Base Titration of Vinegar

	Initial Vol. of NaOH (mL)	Final Vol. of NaOH (mL)	Total Vol. of NaOH (mL)
Trial 1			
Trial 2			
Average			

Show your work for the calculation of the molarity of acetic acid in vinegar.

Show your work for the calculation of the % by mass of acetic acid in vinegar. How does this value compare to the value on the bottle of vinegar?

The Properties of Buffers: Resisting Change in a Turbulent World

D. Van Dinh

OBJECTIVES

- Investigate how buffers work.
- Prepare a buffer solution with a target concentration and pH.
- Determine the buffer capacity of a solution.

INTRODUCTION

Many different systems require the control of conditions so that radical changes do not occur when the system is stressed in some way. For example, our body temperatures remain stable on hot and cold days through perspiration and increased metabolism. In doing so, the effect of the stress due to temperature changes is lessened. In everyday language, a buffer is something that lessens the impact of an external force. One of the most important examples of a system requiring controlled conditions through a buffer is our blood which is a H_2CO_3/HCO_3^- system. Human blood requires a pH between 7.35 and 7.45. If the pH of the blood drops below 6.9 or rises above 7.8, then death is likely. Even within the "safe" range, the optimum pH for many bodily processes is quite narrow; therefore slight disturbances to the pH can significantly impair normal processes. A disturbance that can severely alter the blood pH is cardiac arrest. When the heart stops, metabolic acidosis sets in. Lactic acid and CO_2 cannot be removed from the blood and the pH is dramatically lowered. By applying their knowledge of buffers, hospital emergency room personnel are able to quickly administer the fluids necessary to help restore the victim's blood to its normal state and stave off death (Figure 1). How do they know what concentration of fluids to use in order to restore the blood pH to its normal range?

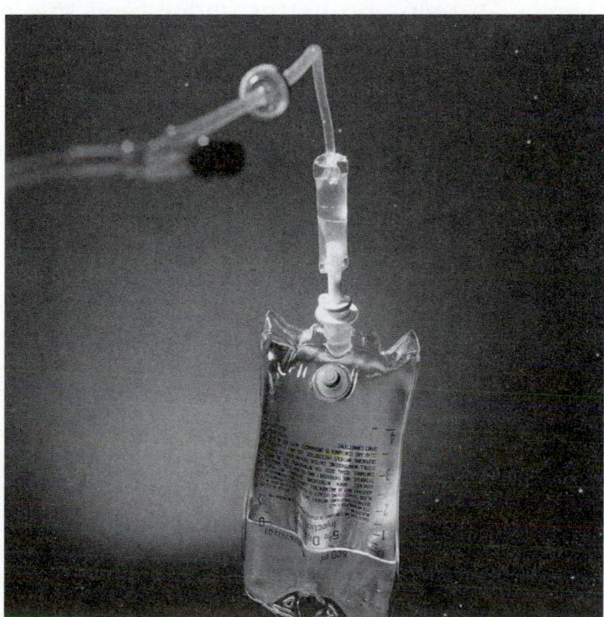

Figure 1. A buffered intravenous drip.

BACKGROUND

The Properties of Buffers

Chemists apply the concept of a buffer to solutions with the ability to resist changes in pH. In many areas of research, chemists need an aqueous solution that resists a pH change when hydrogen ions from a strong acid, or hydroxide ions from a strong base, are added. An acid-base buffer is able to resist changes in its pH by containing an acidic component that can neutralize added OH^- ions and a basic component that can neutralize added H^+ (H_3O^+) ions. Most commonly, the components of a buffer are the conjugate acid-base pair of a weak acid. For example, an aqueous mixture of CH_3COOH and CH_3COO^-. Buffers work through a phenomenon known as the common-ion effect. The essential feature of a buffer is that it consists of high concentrations of the acidic (CH_3COOH) and basic components (CH_3COO^-). This allows the relative concentrations of the buffer components to stay about the same when small amounts of H^+ or OH^- ions are added to the buffer: the acidic component will neutralize any added base and the basic component will neutralize any added acid (Figure 2).

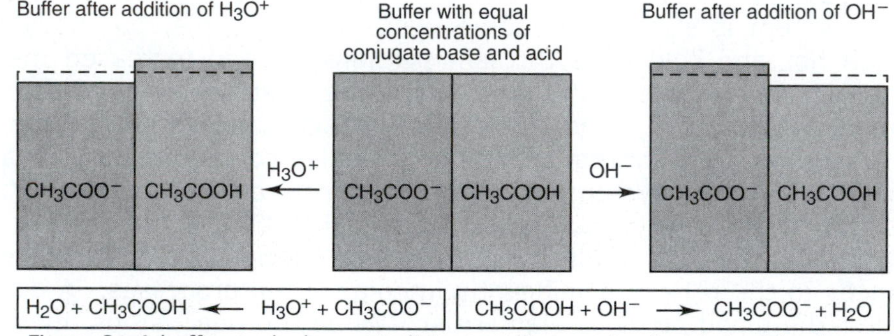

Figure 2. A buffer works by neutralizing small amounts of added acid or base.

It is important to note that a buffer will only work when the amount of H^+ and OH^- added is much smaller than the amounts of the acid-base components of the buffer present. The amount of strong acid or strong base that can be added to a buffer depends on its buffer capacity. The added ions have little effect on the pH because one or the other buffer components consumes them. To help you better understand this concept, let's look at an example.

Consider what happens when small amounts of strong acid or base are added to a buffer containing high $[C_2H_5COOH]$ and $[C_2H_5COO^-]$. As a weak acid, propanoic acid dissociates only slightly in water (Equation 1):

$$C_2H_5COOH\ (aq) \leftrightarrow H^+\ (aq) + C_2H_5COO^-\ (aq) \qquad \text{Equation 1}$$

The expression for the acid dissociation constant is:

$$K_a = \frac{[H^+][C_2H_5COO^-]}{[C_2H_5COOH]} \qquad \text{Equation 2}$$

Noting that K_a is a constant and solving for $[H^+]$, we see that the $[H^+]$ of a solution is dependent on the conjugate acid-base pair concentration ratio:

$$[H^+] = K_a \times \frac{[C_2H_5COOH]}{[C_2H_5COO^-]} \qquad \text{Equation 3}$$

From Equation 3, you can see that if the ratio of the acid to base increases, the $[H^+]$ increases. Likewise, if the ratio of the acid to base decreases, the $[H^+]$ decreases. When a small amount of strong acid is added, such as HCl, the increased amount of H^+ reacts with a nearly stoichiometric amount of $C_2H_5COO^-$ in the buffer to form more C_2H_5COOH (Equation 4).

$$H^+(aq;\ \text{from added HCl}) + C_2H_5COO^-\,(aq) \rightarrow C_2H_5COOH\,(aq) \qquad \text{Equation 4}$$

As a result, the $[C_2H_5COO^-]$ goes down by the amount of added H^+ and the $[C_2H_5COOH]$ goes up by the same amount. This increases the concentration ratio of the acid to base components of the buffer by a slight amount, resulting in a small pH change. Adding a small amount of a strong base, such as NaOH, produces the opposite result.

Preparing a Buffer

As stated in the introduction, many times emergency room personnel are required to administer a buffer solution that simulates the exact concentration and pH of normal human blood. How do they prepare a buffer with a specific concentration and particular pH? In order to calculate the expected pH for a buffer, the Henderson-Hasselbalch equation is used. The Henderson-Hasselbalch equation is a mathematical expression useful for buffer calculations. Taking the negative logarithm of both sides of Equation 3 gives:

$$-\log\,[H^+] = -\log K_a - \log\frac{[C_2H_5COOH]}{[C_2H_5COO^-]} \qquad \text{Equation 5}$$

Substituting definitions of pH and pK_a into Equation 5, we obtain:

$$pH = pK_a + \log\frac{[C_2H_5COO^-]}{[C_2H_5COOH]} \qquad \text{Equation 6}$$

In general, the Henderson-Hasselbalch equation for any conjugate acid-base pair can be written as:

$$pH = pK_a + \log\frac{[\text{base}]}{[\text{acid}]} \qquad \text{Equation 7}$$

In preparing a buffer, the correct $\frac{[\text{base}]}{[\text{acid}]}$ ratio must be determined. Knowing the target pH, one can use the Henderson-Hasselbalch equation in order to calculate the correct concentrations of acid-base ratio to use in a buffer. This ratio gives some idea of the stability of the buffer. The most stable buffers have a ratio of 1:1. This ratio and the target pH are vital to know because it helps us to determine the conjugate acid-base pair to use when preparing a buffer.

For example, imagine you have to prepare 0.250 L of a pH = 5.19 buffer solution of propanoic acid with a total concentration of 0.0500 M from stock solutions of 0.100 M propanoic acid and 0.100 M sodium propanate (NaC_2H_5COO). Propanoic acid (C_2H_5COOH) has a pK_a of 4.89.

To determine the correct $\frac{[C_2H_5COO^-]}{[C_2H_5COOH]}$ ratio, rearrange Equation 6 to obtain:

$$pH - pK_a = \log\frac{[C_2H_5COO^-]}{[C_2H_5COOH]}$$ Equation 8

$$5.19 - 4.89 = 0.30 = \log\frac{[C_2H_5COO^-]}{[C_2H_5COOH]}$$ Equation 8a

Raise both sides to power of 10 to remove log:

$$10^{0.30} = 2.00 = \frac{[C_2H_5COO^-]}{[C_2H_5COOH]}$$ Equation 8b

$$[C_2H_5COO^-] = 2.00[C_2H_5COOH]$$ Equation 8c

This tells us that the amount of $C_2H_5COO^-$ must be twice the amount of C_2H_5COOH. Next, we determine the actual concentrations of the weak acid and the weak base to use in the buffer to meet the required concentration. From Equation 9, we can see that we must have the two buffer component concentrations equal to the total buffer strength.

$$[\text{buffer}] = [\text{acid}] + [\text{base}]$$ Equation 9

Using our example:

$$0.0500\ M = [C_2H_5COOH] + 2.00[C_2H_5COOH]$$ Equation 9a

$$0.0500\ M = 3.00[C_2H_5COOH]$$ Equation 9b

$$0.0167\ M = [C_2H_5COOH]$$ Equation 9c

$$[C_2H_5COO^-] = 2.00[C_2H_5COOH] = 0.0333\ M$$ Equation 9d

The final step in preparing the buffer is to mix the correct volume amounts of the conjugate acid-base pair to give the desired concentration. Knowing the concentrations calculated from Equations 9c-d, the final volume of solution needed (0.250 L) and the initial concentration of the stock solutions (0.100 M) we can use the formula, $M_1V_1 = M_2V_2$, to calculate the exact volume of stock solutions needed.

To find the volume of C_2H_5COOH to use in our buffer we calculate:

$$V_1 = \frac{M_2V_2}{M_1}$$ Equation 10

$$V_1 = \frac{(0.0167\ M)(0.250\ L)}{0.100\ M}$$ Equation 10a

$$V_1 = 0.0418\ L$$ Equation 10b

Repeating these steps to calculate the volume of $C_2H_5COO^-$ to use would give us 0.0833 L. Thus, we mix 41.8 mL of 0.100 M propanoic acid solution and 83.3 mL of 0.100 M sodium propanate solution in a 250.0-mL volumetric flask. Adding de-ionized H_2O while mixing will give a total solution with a volume of 250.0-mL, a pH of 5.19 and a total concentration of 0.0500 M.

OVERVIEW

In this lab you will learn how to prepare a buffer of a desired pH and concentration. You will also observe how buffers stabilize the pH of a solution even after adding a strong acid or strong base. In Part A, you will be acquainted with the procedures and calculations involved in preparing a buffer. You will prepare one of three buffer solutions of ammonia/ammonium chloride with varying pH. You will test this buffer with a strong acid and strong base and exchange data with other groups. Using the knowledge gained from Part A, you will then prepare your own buffer in Part B, given a total buffer concentration, a target pH, and a choice of two conjugate acid-base pairs. In Part C, you will titrate the buffer you made in Part B with a strong base in order to determine its buffer capacity.

PROCEDURE

Part A. Properties of a Buffer

Chemicals Used	Materials Used
0.10 M H_2CO_3l	pH Meter
0.10 M $NaHCO_3$	Graduated pipet and bulb
0.50 M HCl	50-mL Graduated cylinder
0.50 M NaOH	100-mL Volumetric flask
	100-mL Beaker (4)
	Funnel
	Plastic pasteur pipet

You will be assigned to prepare one of the following three $H_2CO_3/NaHCO_3$ buffer solutions. Carbonic acid has a pK_a of 6.37.

Buffer 1: Prepare 0.100 L of a pH = 6.37 buffer with a total concentration of 0.0400 M by mixing 20.0 mL of 0.10 M H_2CO_3 with 20.0 mL of 0.10 M $NaHCO_3$.

Buffer 2: Prepare 0.100 L of a pH = 6.85 buffer with a total concentration of 0.0400 M by mixing 10.0 mL of 0.10 M H_2CO_3 with 30.0 mL of 0.10 M $NaHCO_3$.

Buffer 3: Prepare 0.100 L of a pH = 5.89 buffer with a total concentration of 0.0400 M by mixing 30.0 mL of 0.10 M H_2CO_3 with 10.0 mL of 0.10 M $NaHCO_3$.

1. After being assigned a buffer, prepare 0.100 L of the buffer in a 100-mL volumetric flask using the materials listed above.

2. Using a clean, dry 50-mL graduated cylinder, measure two 25.0-mL portions of your buffer solution and pour into two separate 100-mL beakers. In two other 100-mL beakers, measure out two 25.0-mL portions of de-ionized H_2O.

3. With a pH meter, record the initial pH of each buffer solution and each sample of de-ionized H_2O. Be sure to rinse the tip of the pH meter with de-ionized H_2O and dry it between uses. With a clean, disposable pipet, add 5 drops of 0.50 M HCl to one of the beakers containing your buffered solution and 5 drops of 0.50 M NaOH to the other beaker containing your buffered solution. Stir the solutions and record the resulting pH. Repeat this procedure with the two samples of de-ionized H_2O. Exchange data with groups who made the other two buffer solutions.

4. Answer the "Questions from Part A" before performing Part B of the experiment.

PART B. Designing a Buffer

Chemicals Used	Materials Used
0.50 M $HC_2H_3O_2$ 0.50 M $NaC_2H_3O_2$ 0.50 M HCOOH 0.50 M NaCOOH	Graduated pipet and bulb 100-mL Volumetric flask Funnel

You will be assigned to prepare one of the following four buffer solutions.

Buffer 1: Prepare 0.100 L of a pH = 3.74 buffer with a total concentration of 0.100 M.

Buffer 2: Prepare 0.100 L of a pH = 4.04 buffer with a total concentration of 0.100 M.

Buffer 3: Prepare 0.100 L of a pH = 4.44 buffer with a total concentration of 0.100 M.

Buffer 4: Prepare 0.100 L of a pH = 5.04 buffer with a total concentration of 0.100 M.

After being assigned a buffer system to make, choose the appropriate conjugate acid-base pair (either $HC_2H_3O_2/NaC_2H_3O_2$ or HCOOH/NaCOOH) to use. Calculate the appropriate volumes to be used and make your buffer. Be sure to save your buffer for Part C.

PART C. Determination of Buffer Capacity

Chemicals Used	Materials Used
0.50 M NaOH Buffer from Part B (50 mL)	pH Meter 50-mL Graduated cylinder 100-mL, 250-mL Beakers 50-mL Buret and buret stand Funnel

1. Prepare a 50-mL buret by running about 25.0 mL of de-ionized H_2O through it and then rinsing the buret with about 5.0 mL of 0.50 M NaOH solution. Mount the buret in the stand. Using a funnel, pour 25.0 mL of the 0.50 M NaOH solution into the buret. Drain any air bubbles from the tip of the buret into a waste beaker. Record the initial volume.

2. Pour 50.0 mL of the buffer you made in Part B into a 250-mL beaker. Titrate your buffer by adding the 0.50 M NaOH solution drop-wise. While monitoring the pH, record the volume of NaOH needed to reach your buffer's capacity.

3. Exchange data with other groups who had a different buffer.

The Properties of Buffers: Resisting Change in a Turbulent World

Name:	Lab Instructor:
Date:	Lab Section:

PRE-LABORATORY EXERCISES

1. Define the term "common-ion effect" used in the BACKGROUND section. How do buffers work by taking advantage of the common-ion effect?

2. Give an example of a salt that could be used to make a buffer with HCN.

3. If Solution A has a pH of 3.23 and Solution B has a pH of 4.23, what is their relationship in terms of $[H^+]$?

4. What is the relationship between the concentration of buffer components and the buffer capacity? Explain.

5. For a buffer having a $\frac{[\text{base}]}{[\text{acid}]}$ ratio of 1:1, what is the relationship between pH and pK_a?

OVER →

PRE-LABORATORY EXERCISES continued...

6. Using Equations 1 and 2 as models, write the acid dissociation reaction and the K_a expression for the acetic and formic acid used in Part B. Find the K_a values for acetic and formic acid and calculate their pK_a values (also record these values in your laboratory notebook so you will have them available during the experiment).

The Properties of Buffers: Resisting Change in a Turbulent World

Name:	Lab Instructor:
Date:	Lab Section:

RESULTS and POST-LABORATORY QUESTIONS

Part A. Properties of a Buffer

Which set of data is yours? Buffer 1, Buffer 2 or Buffer 3

	Buffer 1	Buffer 2	Buffer 3
$\frac{[\text{base}]}{[\text{acid}]}$ ratio			
Average pH of buffer			
Average pH of DI-H_2O			
pH of buffer after adding HCl			
pH of buffer after adding NaOH			
pH of DI-H_2O after adding HCl			
pH of DI-H_2O after adding NaOH			

QUESTIONS FROM PART A

1. Verify that the volumes of acid and base you were asked to use will result in the correct total concentration of the buffer.

2. Calculate the $\frac{[\text{base}]}{[\text{acid}]}$ ratio used for your buffer.

3. Closely examine the data from Part A. Looking at all three buffer solutions, do you see a relationship between the pK_a value of the acid, the desired or target pH of the buffer, and the $\frac{[\text{base}]}{[\text{acid}]}$ ratio? Explain.

OVER →

RESULTS and POST-LABORATORY QUESTIONS continued...

4. How does the relationship you found in QUESTION 3 help you decide which conjugate acid-base pair to choose when designing a buffer?

5. After performing Part A, did you notice a difference between the buffer and the DI-water when the strong acid/base was added? Did the buffer "do it's job"? Explain.

PART B. Designing a Buffer

Attach a copy of your work showing the calculations of the volumes of acid and base needed to create the target buffer you were assigned.

PART C. Determination of Buffer Capacity

Which set of data is yours? Buffer 1, Buffer 2, Buffer 3 or Buffer 4

	Buffer 1	**Buffer 2**	**Buffer 3**	**Buffer 4**
Conjugate acid-base pair used				
Moles of acid in buffer				
Total vol. of NaOH to reach buffer capacity				
Total moles of NaOH to reach buffer capacity				

How does the total moles of NaOH needed to reach buffer capacity relate to the number of moles of acid in the buffer? Explain.

Electrochemistry: An Introduction to Voltaic Cells

David Roberts

OBJECTIVES

- Define basic electrochemistry terms and concepts.
- Construct an activity series by ranking the relative reactivity of several metals.
- Build a working voltaic cell.
- Experimentally determine the electrical potential for your voltaic cell.

INTRODUCTION

Electrochemistry is life. The nerve impulses in biological systems exist due to electrical chemical potentials. It is these potentials that enable our eyes to sense light, our heart to beat, our blood to transport oxygen and our mind to understand. You can actually see your body's electrochemical potential by grasping both ends of a voltmeter.

Electrochemical processes are crucial, not only to life, but also in the pursuit of a comfortable livelihood. Advances in electrochemistry have brought us batteries for flashlights, lithium cells for pacemakers, and lead acid batteries for automobiles (Figure 1). Electrochemistry has made modern day life more comfortable and efficient. Can you think of an electrochemical device that makes your life easier?

Fig 1. The electrochemical energy provided by batteries comes from many varied chemical reactions.

BACKGROUND

Since atoms are bound together by electrons, the fundamental units of electricity and magnetism, it shouldn't be surprising that there is a close link between some chemical reactions and electricity. The field of chemistry involving the study of the overlap between these two areas is called electrochemistry or redox (**ox**idation/**red**uction) chemistry. The defining characteristic of redox reactions is the transfer of electrons from one species to another. The reaction between zinc metal and lead ions is a typical example that illustrates this characteristic (Equation 1).

$$Zn\ (s) + Pb^{2+}\ (aq) \leftrightarrow Zn^{2+}\ (aq) + Pb\ (s) \qquad \text{Equation 1}$$

The oxidation of zinc metal and the reduction of lead ions is due to the relative activity of the two elements. The more active the metal; the more likely it is to be oxidized and serve as a reducing agent. A less active species is more likely to be reduced and serve as an oxidizing agent. In Equation 1, zinc metal is more active than lead and, therefore, is oxidized. This concept of activity is the driving force of spontaneous electrochemistry. If you refer to an activity series, you will find that the more active the species the greater its tendency to lose electrons. The activity of a species depends upon its ionization energy. Small ionization energies correspond to more active metals. Table 1 illustrates these relationships. The table is composed of a series of redox reactions in which the strongest oxidizing agent (reactant) and the weakest reducing agent (product) is located towards the top of the chart. As you proceed down the chart the strength of the oxidizing agent decreases and the strength of the reducing agent increases. This trend allows you to determine by inspection the likelihood for a species to be reduced in comparison to another. In an iron-copper reaction, for instance, Cu^{2+} ion is the stronger oxidizing agent in comparison to Fe^{2+} and, therefore, is reduced.

Table 1. Standard-State Reduction Half-Cell Potentials

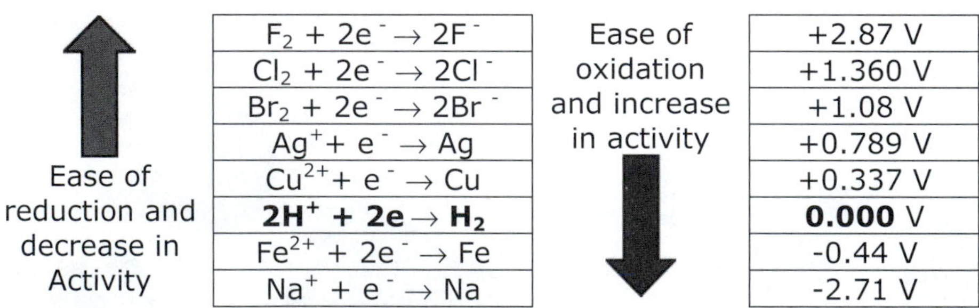

$F_2 + 2e^- \rightarrow 2F^-$	+2.87 V
$Cl_2 + 2e^- \rightarrow 2Cl^-$	+1.360 V
$Br_2 + 2e^- \rightarrow 2Br^-$	+1.08 V
$Ag^+ + e^- \rightarrow Ag$	+0.789 V
$Cu^{2+} + e^- \rightarrow Cu$	+0.337 V
$2H^+ + 2e \rightarrow H_2$	**0.000** V
$Fe^{2+} + 2e^- \rightarrow Fe$	-0.44 V
$Na^+ + e^- \rightarrow Na$	-2.71 V

When you are able to use Table 1 to determine the activity and oxidation/reduction strength of a species, you are on your way to being able to understand the principles that effect electrochemical systems such as voltaic cells. Voltaic cells harness the energy that is released during a redox reaction by separating the redox process into two distinct half reactions. The separation of these reactions generates the flow of electrons when connected together by an external source such as a wire.

Voltaic Cells

A voltaic cell is an electrochemical system that converts chemical energy to electrical energy. The energy is produced by a spontaneous redox reaction within a cell. Figure 2 shows a voltaic cell. It is divided into two half-cells which each contain one half of the total redox reaction. This voltaic cell makes use of a zinc/copper reaction. In a half-cell the zinc electrode is immersed in a solution of zinc sulfate while the copper electrode is immersed in a copper sulfate solution. The electrodes in the voltaic cells are conductors that will receive or donate electrons. If the electrode receives electrons from the external circuit then it will be labelled the cathode. If it donates electrons through the external circuit then it will be labelled the anode. A salt bridge separates these solutions.

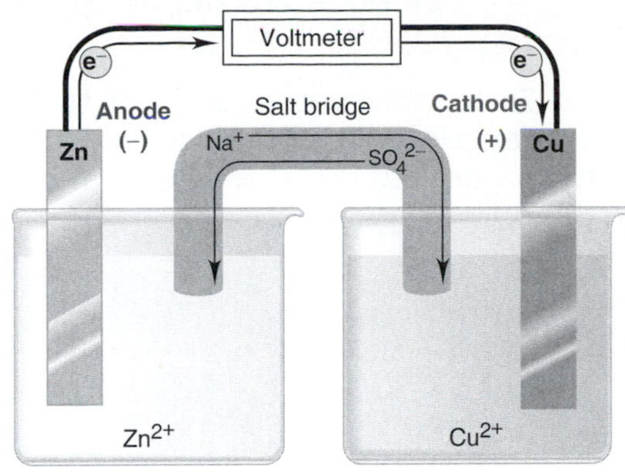

Figure 2. A copper and zinc voltaic cell

Oxidation half-reaction:	$Zn\ (s) \rightarrow Zn^{2+}\ (aq) + 2e^-$	Equation 2
Reduction half-reaction:	$2e^- + Cu^{2+}\ (aq) \rightarrow Cu\ (s)$	Equation 2a
Overall reaction:	$Zn\ (s) + Cu^{+2}\ (aq) \rightarrow Zn^{+2}\ (aq) + Cu\ (s)$	Equation 2b

In Figure 2, electrons spontaneously flow in a clockwise direction, from the zinc electrode to the copper electrode and a voltage is produced. This voltage is determined by inserting a voltmeter into the system that measures the reduction potential difference between the cathode and the anode. This difference is due to the reactions that are occurring at these electrodes.

Calculating Electrical Potential

We can calculate the voltage in any cell by taking the difference in reduction potentials between the cathode and the anode (Equation 3).

$$E^\circ_{Cell} = E^\circ_{Cathode} - E^\circ_{Anode} \qquad \text{Equation 3}$$

Although tables of standard cell potential differences based on Equation 3 would be helpful, it is more useful to have a table of standard half-cell reactions. This is because this type of table gives more information in a lot less space. Tables of standard electrode potentials can be obtained if any one electrode, operated under standard conditions, is designated as the standard electrode or standard reference electrode with which all other electrodes will be compared. This standard reference electrode can then arbitrarily be assigned an electrode potential of zero, just as sea level is assigned the elevation of zero. Chemists have agreed to select the electrode reaction of hydrogen, as the standard half-reaction against which all others will be compared. This electrode is called the standard hydrogen electrode. The use of the standard hydrogen electrode involves hydrogen gas being turned into hydrogen ions. During this process the hydrogen electrode is conventionally placed at the cathode of a cell and the potential of this half-cell is given the value of 0 V. The potentials for other cells are then tabulated relative to this value. For example, if we make a voltaic cell comparing Zn to the standard hydrogen electrode, we measure a cell potential of 0.76 V. We can use this information to determine the value of the half-cell potential for zinc. We begin by writing the half-cell reactions (Equations 4 and 4a) for the overall cell (Equation 4b).

Oxidation half-reaction (Anode):	$Zn\ (s) \rightarrow Zn^{2+}\ (aq) + 2e^-$	Equation 4
Reduction half-reaction (Cathode):	$2H^+\ (aq) + 2e^- \rightarrow H_2\ (g)$	Equation 4a
Overall reaction:	$Zn\ (s) + 2H^+\ (aq) \rightarrow Zn^{2+}\ (aq) + H_2\ (g)$	Equation 4b

The potential of this cell is 0.76 V. However, this is not the standard reduction potential of zinc. We must calculate the standard reduction potential by following Equation 3. Given that the electrode potential of hydrogen is 0.00 V, we now can solve for the standard reduction potential for zinc by using Equation 3.

$$E^\circ_{Cell} = E^\circ_{Cathode} - E^\circ_{Anode} \qquad \text{Equation 3}$$

$$0.76\ V = 0.00\ V - E^\circ_{Anode} \qquad \text{Equation 3a}$$

$$E^\circ_{Anode} = E^\circ_{Zn} = -\ 0.76\ V \qquad \text{Equation 3b}$$

Using Equation 3 and a standard reference electrode, potentials can be obtained for any species. Table 1 and the table of Standard Reduction Potentials in your text were created relative to hydrogen. In this laboratory experiment you will make your own standard reduction potential table using copper as your reference electrode.

OVERVIEW

In Part A of this laboratory experiment, you will qualitatively determine the relative reactivity of copper, zinc and nickel. In Part B, you will make quantitative observations of the reactivity of these metals by measuring the reduction potential of various voltaic cells.

PROCEDURE

Part A. Qualitative determination of relative reactivity of the metals

Chemicals Used	Materials Used
Cu, Ni, Zn metal wire 1.00 M solutions of Cu, Ni and Zn nitrates	24-Well plate 50-mL Beaker (4) 250-mL Beaker for waste Disposable pipets Wire cutters Tweezers Steel wool

Clean all metal pieces with steel wool to remove oxide coatings and then cut to 1 cm length. Using the provided materials, devise a way in which you can determine the relative reactivity of the metals listed above. Dispose of solutions according to your instructor's directions.

Part B. Determination of Reduction Potentials

Chemicals Used	Materials Used
Cu, Ni, Zn, Ag metals 1.00 M solutions of Cu, Ni, Zn, Ag, Na nitrate	50-mL Beaker (2) 250-ml Beaker for waste Scissors Filter paper Wire cutters Voltmeter Steel wool

Each pair of students will be assigned to build one or two voltaic cells using different combinations of Cu, Ni, Zn, and Ag. Working in pairs, construct your assigned voltaic cell. Each person in a pair should make one of the half-cells. Begin by cleaning your metal wire with steel wool. Cut a length of wire approximately 2.5 inches in length to serve as your electrode. Half-fill a 50-ml beaker with your half-cell solution. The salt bridge is prepared by cutting a section of filter paper into strips. Saturate the strip of filter paper with sodium nitrate and use it to bridge the two half-cells. Make sure the salt bridge touches the solution in both half-cells and that it does not dry out during the experiment. After you measure the potential for your cell, move to other stations until you have measured all 6 possible cells. Dispose of cell contents according to your instructor's directions.

Electrochemistry: An Introduction to Voltaic Cells

Name:	Lab Instructor:
Date:	Lab Section:

PRE-LABORATORY EXCERCISES

1. Define all <u>underlined</u> terms in the BACKGROUND section of this lab.

2. Write the reduction half-reaction for each of the species in the following pairs. Based on the standard reduction potential for each species, calculate E_{cell} for the electrochemical cell made from each pair of species.
   a) Ag and Sn

   b) Al and Ni

   c) F_2 and Mg

OVER →

PRE-LABORATORY EXCERCISES continued...

3. Draw and label the major components of a voltaic cell. Describe the role of each of the components. Be sure to compare and contrast the chemical processes that occur at the cathode and the anode.

Electrochemistry: An Introduction to Voltaic Cells

Name:	Lab Instructor:
Date:	Lab Section:

RESULTS and POST-LABORATORY QUESTIONS

Part A. Qualitative determination of relative reactivity of the metals

Fully explain your approach, in Part A, to ranking Cu, Ni, and Zn from the least to the most active. What are your results?

Part B. Determination of Reduction Potentials

Cell	Measured E_{cell}
Ag and Cu	
Zn and Cu	
Ni and Cu	

Using the above measured cell potentials, fill in the following chart to create a standard reduction table based on copper. Assign copper a value of 0.00 V.

Half Cell Reaction	E°(v)

Now that you have completed the standard reduction table, complete the following chart:

Cell	Measured E_{cell}	E_{cell} related to Cu (use table above)	E_{cell} related to H_2 (use text)
Ni and Zn			
Ag and Zn			
Ag and Ni			

OVER →

RESULTS and POST-LABORATORY QUESTIONS continued...

1. Show your work for all of the calculations in the previous table.

2. What is the relationship the between "E_{cell} values related to Cu" and the "E_{cell} related to H_2" in the above table?

3. Do the results from Part B confirm or contradict the qualitative ranking from Part A? Explain.

4. Using your data from Part B where does Ag fit into the ranking from Part A?

Credits:

Chapter 1
p.1-1: Courtesy Jeffrey Paradis

Chapter 2
p. 2-1: NASA; p. 2-2: © McGraw-Hill Higher Education/Stephen Frisch Photographer; p. 2-3: © Getty Images/PhotoDisc/Vol. 34/Spacescapes

Chapter 3
p. 3-1: © Bridgeman-Giraudon/Art Resource, NY

Chapter 4
p. 4-1: © Werner H. Muller/Peter Arnold, Inc.; p. 4-3: © McGraw-Hill Higher Education/Stephen Frisch Photographer

Chapter 5
p. 5-1: © Getty Images/Photodisc/Vol. 72 Professional Science; p. 5-3: Courtesy of Mettler; p. 5-4 top left: © Corbis/Vol. 154/Doctors & Medicine; p. 5-4 top right, bottom left, bottom right: Courtesy Jeffrey Paradis; p. 5-5 top left, top middle, top right, bottom right, bottom left: Courtesy Jeffrey Paradis

Chapter 7
p. 7-2: © McGraw-Hill Higher Education/Stephen Frisch Photographer; p. 7-4: Courtesy Jeffrey Paradis

Chapter 8
p. 8-1: © Corbis/Modern Cuisine Vol. 43; p. 8-6: Courtesy Jeffrey Paradis

Chapter 9
p. 9-1: © Richard Kaylin/Tony Stone Images/Getty; p. 9-3: © McGraw-Hill Higher Education/Stephen Frisch Photographer

Chapter 10
p. 10-2: © McGraw-Hill Higher Education/Stephen Frisch Photographer

Chapter 11
p. 11-1: © Getty Images/Photodisc/Douglas Miller; p. 11-2: ©Corbis/Vol. 45/Destination Africa; p. 11-3: © McGraw-Hill Higher Education/Stephen Frisch Photographer; p. 11-5: Courtesy Jeffrey Paradis

Chapter 12
p. 12-1: Courtesy Mike Ware

Chapter 13
p. 13-1: © Andrew Brookes/CORBIS; p. 13-2: Courtesy Betz Corp.

Chapter 14
p. 14-6: © McGraw-Hill Higher Education/Stephen Frisch Photographer

Chapter 15
p. 15-2 all: © McGraw-Hill Higher Education/Stephen Frisch; p. 15-5: Courtesy Jeffrey Paradis

Chapter 16
p. 16-1: © Romilly Lockyer/Image Bank/Getty Images

Chapter 17
p. 17-1, 17-2: Courtesy Jeffrey Paradis

Chapter 18
p. 18-4, 18-8 both: Courtesy Jeffrey Paradis

Chapter 22
p. 22-1 left: © D. Osf/Animals Animals/Earth Scenes; p. 22-1 right: © Corbis/Vol. 135/Family Time 2

Chapter 24
p. 24-1: Courtesy World Meteorlogical Organization and Laboratory of Atmospheric Physics, Aristotle University of Thessaloniki, Greece (LAP-AUTH-GR 1999)

Chapter 25
p. 25-1: © Corbis/Vol. 116 Children at Play; p. 25-2 all: © McGraw-Hill Higher Education/Stephen Frisch Photographer

Chapter 26
p. 26-1: © Kennan Ward/2004

Chapter 27
p. 27-1: © Getty Images/David Perry; p. 27-5: © McGraw-Hill Higher Education/Stephen Frisch Photographer

Chapter 28
p. 28-1: © Getty Images/Michael Matisse

Chapter 29
p. 29-1: © Getty Images/Photodisc

Notes:

Notes: